IET CONTROL, ROBOTICS AND SENSORS SERIES 134

Human-Robot Collaboration

Other volumes in this series:

Volume 8	**A History of Control Engineering, 1800–1930** S. Bennett
Volume 9	**Embedded Mechatronics System Design for Uncertain Environments: Linux-based, MATLAB xPC Target, PIC, Ardunio and Raspberry Pi Approaches** C.S. Chin
Volume 18	**Applied Control Theory, 2nd Edition** J.R. Leigh
Volume 20	**Design of Modern Control Systems** D.J. Bell, P.A. Cook and N. Munro (Editors)
Volume 28	**Robots and Automated Manufacture** J. Billingsley (Editor)
Volume 32	**Human Machine Collaboration and Interaction for Smart Manufacturing: Automation, robotics, sensing, artificial intelligence, 5G, IoTs and Blockchain** W.Y. Leong (Editor)
Volume 33	**Temperature Measurement and Control** J.R. Leigh
Volume 34	**Singular Perturbation Methodology in Control Systems** D.S. Naidu
Volume 35	**Implementation of Self-tuning Controllers** K. Warwick (Editor)
Volume 37	**Industrial Digital Control Systems, 2nd Edition** K. Warwick and D. Rees (Editors)
Volume 39	**Continuous Time Controller Design** R. Balasubramanian
Volume 40	**Deterministic Control of Uncertain Systems** A.S.I. Zinober (Editor)
Volume 41	**Computer Control of Real-time Processes** S. Bennett and G.S. Virk (Editors)
Volume 42	**Digital Signal Processing: Principles, devices and applications** N.B. Jones and J.D.McK. Watson (Editors)
Volume 44	**Knowledge-based Systems for Industrial Control** J. McGhee, M.J. Grimble and A. Mowforth (Editors)
Volume 47	**A History of Control Engineering, 1930–1956** S. Bennett
Volume 49	**Polynomial Methods in Optimal Control and Filtering** K.J. Hunt (Editor)
Volume 50	**Programming Industrial Control Systems Using IEC 1131-3** R.W. Lewis
Volume 51	**Advanced Robotics and Intelligent Machines** J.O. Gray and D.G. Caldwell (Editors)
Volume 52	**Adaptive Prediction and Predictive Control** P.P. Kanjilal
Volume 53	**Neural Network Applications in Control** G.W. Irwin, K. Warwick and K.J. Hunt (Editors)
Volume 54	**Control Engineering Solutions: A practical approach** P. Albertos, R. Strietzel and N. Mort (Editors)
Volume 55	**Genetic Algorithms in Engineering Systems** A.M.S. Zalzala and P.J. Fleming (Editors)
Volume 56	**Symbolic Methods in Control System Analysis and Design** N. Munro (Editor)
Volume 57	**Flight Control Systems** R.W. Pratt (Editor)
Volume 58	**Power-plant Control and Instrumentation: The control of boilers and HRSG systems** D. Lindsley
Volume 59	**Modelling Control Systems Using IEC 61499** R. Lewis
Volume 60	**People in Control: Human factors in control room design** J. Noyes and M. Bransby (Editors)
Volume 61	**Nonlinear Predictive Control: Theory and practice** B. Kouvaritakis and M. Cannon (Editors)
Volume 62	**Active Sound and Vibration Control** M.O. Tokhi and S.M. Veres
Volume 63	**Stepping Motors, 4th Edition** P.P. Acarnley
Volume 64	**Control Theory, 2nd Edition** J.R. Leigh
Volume 65	**Modelling and Parameter Estimation of Dynamic Systems** J.R. Raol, G. Girija and J. Singh
Volume 66	**Variable Structure Systems: From principles to implementation** A. Sabanovic, L. Fridman and S. Spurgeon (Editors)
Volume 67	**Motion Vision: Design of compact motion sensing solution for autonomous systems** J. Kolodko and L. Vlacic
Volume 68	**Flexible Robot Manipulators: Modelling, simulation and control** M.O. Tokhi and A.K.M. Azad (Editors)
Volume 69	**Advances in Unmanned Marine Vehicles** G. Roberts and R. Sutton (Editors)
Volume 70	**Intelligent Control Systems Using Computational Intelligence Techniques** A. Ruano (Editor)
Volume 71	**Advances in Cognitive Systems** S. Nefti and J. Gray (Editors)
Volume 72	**Control Theory: A guided tour, 3rd Edition** J. R. Leigh
Volume 73	**Adaptive Sampling with Mobile WSN** K. Sreenath, M.F. Mysorewala, D.O. Popa and F.L. Lewis
Volume 74	**Eigenstructure Control Algorithms: Applications to aircraft/rotorcraft handling qualities design** S. Srinathkumar
Volume 75	**Advanced Control for Constrained Processes and Systems** F. Garelli, R.J. Mantz and H. De Battista
Volume 76	**Developments in Control Theory towards Glocal Control** L. Qiu, J. Chen, T. Iwasaki and H. Fujioka (Editors)
Volume 77	**Further Advances in Unmanned Marine Vehicles** G.N. Roberts and R. Sutton (Editors)
Volume 78	**Frequency-Domain Control Design for High-Performance Systems** J. O'Brien
Volume 80	**Control-oriented Modelling and Identification: Theory and practice** M. Lovera (Editor)
Volume 81	**Optimal Adaptive Control and Differential Games by Reinforcement Learning Principles** D. Vrabie, K. Vamvoudakis and F. Lewis
Volume 83	**Robust and Adaptive Model Predictive Control of Nonlinear Systems** M. Guay, V. Adetola and D. DeHaan
Volume 84	**Nonlinear and Adaptive Control Systems** Z. Ding
Volume 86	**Modeling and Control of Flexible Robot Manipulators, 2nd edition** M. O. Tokhi and A.K.M. Azad
Volume 88	**Distributed Control and Filtering for Industrial Systems** M. Mahmoud
Volume 89	**Control-based Operating System Design** A. Leva *et al.*
Volume 90	**Application of Dimensional Analysis in Systems Modelling and Control Design** P. Balaguer
Volume 91	**An Introduction to Fractional Control** D. Valério and J. Costa
Volume 92	**Handbook of Vehicle Suspension Control Systems** H. Liu, H. Gao and P. Li

Volume 93 **Design and Development of Multi-Lane Smart Electromechanical Actuators** F.Y. Annaz
Volume 94 **Analysis and Design of Reset Control Systems** Y. Guo, L. Xie and Y. Wang
Volume 95 **Modelling Control Systems Using IEC 61499, 2nd Edition** R. Lewis & A. Zoitl
Volume 96 **Cyber-Physical System Design with Sensor Networking Technologies** S. Zeadally and N. Jabeur (Editors)
Volume 97 **Sensory Systems for Robotic** Applications R. Dahiya O. Gordon and C. Ozioko (Editors)
Volume 99 **Practical Robotics and Mechatronics: Marine, Space and Medical Applications** I. Yamamoto
Volume 100 **Organic Sensors: Materials and Applications** E Garcia-Breijo and P Cosseddu (Editors)
Volume 102 **Recent Trends in Sliding Mode Control** L. Fridman JP. Barbot and F. Plestan (Editors)
Volume 104 **Control of Mechatronic Systems** L. Guvenc, B.A Guvenc, B. Demirel, M.T Emirler
Volume 105 **Mechatronic Hands: Prosthetic and Robotic Design** P.H. Chappell
Volume 107 **Solved Problems in Dynamical Systems and Control** D. Valério, J. T Machado, A. M. Lopes and A. M. Galhano
Volume 108 **Wearable Exoskeleton Systems: Design, Control and Applications** S. Bai, G.S. Virk and T.G.Sugar
Volume 111 **The Inverted Pendulum in Control Theory and Robotics: From Theory to New Innovations** O. Boubaker and R. Iriarte (Editors)
Volume 112 **RFID Protocol Design, Optimization, and Security for the Internet of Things** A. X. Liu, M. Shahzad, X Liu and K. Li
Volume 113 **Design of Embedded Robust Control Systems Using MATLAB®/Simulink®** P.H. Petkov, T.N. Slavov and J.K. Kralev
Volume 114 **Signal Processing and Machine Learning for Brain-Machine Interfaces** T. Tanaka and M. Arvaneh (Editor)
Volume 116 **Imaging Sensor Technologies and Applications** W Yang (Editor)
Volume 117 **Data Fusion in Wireless Sensor Networks** D. Ciuonzo and P.S. Rossi (Editors)
Volume 118 **Modeling, Simulation and Control of Electrical Drives** M. F. Rahman, S. K. Dwivedi (Editors)
Volume 119 **Swarm Intelligence Volumes 1–3** Y. Tan (Editor)
Volume 120 **Imaging and Sensing for Unmanned Aircraft Systems Vol1 and 2** V. Estrela, J. Hemanth, O. Saotome, G. Nikolakopoulos and R. Sabatini (Editors)
Volume 121 **Integrated Fault Diagnosis and Control Design of Linear Complex Systems** M. Davoodi, N. Meskin and K. Khorasani
Volume 123 **Data-Driven Modeling, Filtering and Control: Methods and applications** C. Novara and S. Formentin (Editors)
Volume 125 **Short-Range Micro-Motion Sensing with Radar Technology** C. Gu and J.Lien (Editors)
Volume 126 **Fault Diagnosis and Fault-tolerant Control of Robotic and Autonomous Systems** Monteriù, Longhi and Freddi (Editors)
Volume 127 **Sensors, Actuators, and Their Interfaces: A multidisciplinary introduction, 2nd Edition** N. Ida
Volume 128 **IoT Technologies in Smart Cities: From sensors to big data, security and trust** F. Al-Turjman and M. Imran (Editors)
Volume 129 **Signal Processing to Drive Human-Computer Interaction: EEG and eye-controlled interfaces** I. Kompatsiaris, C. Kumar and S. Nikolopoulos (Editors)
Volume 130 **Transparency for Robots and Autonomous Systems: Fundamentals, technologies and applications** R. Wortham
Volume 131 **Space Robotics and Autonomous Systems: Technologies, advances and applications** Yang Gao (Editor)

Human-Robot Collaboration

Unlocking the potential for industrial applications

Edited by
Zoe Doulgeri and Fotios Dimeas

The Institution of Engineering and Technology

Published by The Institution of Engineering and Technology, London, United Kingdom

The Institution of Engineering and Technology is registered as a Charity in England & Wales (no. 211014) and Scotland (no. SC038698).

First published 2023

The Institution of Engineering and Technology
Futures Place
Kings Way, Stevenage
Herts, SG1 2UA, United Kingdom

www.theiet.org

British Library Cataloguing in Publication Data
A catalogue record for this product is available from the British Library

ISBN 978-1-83953-598-7 (hardback)
ISBN 978-1-83953-599-4 (PDF)

Typeset in India by Exeter Premedia Services Private Limited

Cover credit: Jacopo Guastamacchia/A human collaborating with an industrial robot

Contents

About the Editors

Zoe Doulgeri is a professor of Robotics and Control of Manufacturing Systems and the director of the Automation and Robotics Lab of the Department of Electrical and Computer Engineering of the Aristotle University of Thessaloniki (AUTH). She received a diploma of Electrical Engineering from AUTH and an MSc (DIC) in Control Systems, an MSc (DIC) in Social and Economic Aspects of Science and Technology in Industry, and a PhD in Mechanical Engineering from the Imperial College, London, UK. She teaches Control Systems and Robotics. She authored more than 150 publications in peer-reviewed international journals and conferences. She is an evaluator of EU research and innovation proposals in Horizon Europe and a reviewer in the most important robotics journals and conferences. She served as associate editor in the Journal of Intelligent and Robotic Systems and the IEEE Robotics and Automation Letters. She has participated in many robotics projects financed by the EU and the Greek government. She has been the coordinator of the recently completed H2020 "Collaborate" project. She is currently coordinating the H2020 European research project BACCHUS on mobile robotic platforms for active inspection and harvesting in agricultural areas and two national projects. Her current research interests include the topics of physical human-robot interaction, robot teaching and learning by demonstration, bimanual mobile robots, object grasping and manipulation with analytical and data-based learning methods and the control of uncertain robotic systems. She has been the representative of Greece in the European Control Association from 2016 to 2021. She is a senior member of the IEEE, the IEEE Robotics and Automation Society, and the IEEE Control Systems Society.

Fotios Dimeas is a senior research fellow in the Automation and Robotics Lab at AUTH, Greece. He received his master's degree in Mechanical Engineering in 2012 and his PhD degree in Robotics in 2016 from the Department of Mechanical Engineering and Aeronautics of the University of Patras, Greece. His research covers human-robot collaboration, including teaching by demonstration, motion generation and robot control. During his PhD and PostDoc studies, he received multiple scholarships from the National Scholarship Foundation and the National Education and Lifelong Learning Programme. He has authored more than 30 publications in peer-reviewed conferences and journals, including *ICRA*, *IROS*, *SMC*, *IEEE Transaction in Haptics, Robotics and Computer-Integrated Manufacturing*, *IEEE Robots and Automation Letters*, and *IEEE Robots and Automation Magazine*. He has participated as a researcher in the Horizon 2020 projects SARAFun, CoLLaboratE and BACCHUS and as a technical coordinator in the national project ProgHRC.

Foreword

Human-robot collaboration (HRC) has emerged as a prominent research field in recent years with the aim to revolutionize the industrial robotics automation landscape by introducing a human in the loop. However, despite the progress and the increasing availability of collaborative robots, their adoption rate from the industry is low and is usually limited to basic workspace sharing. The need to unlock the potential of robots for actual collaborative work, where robots can operate intelligently and adaptively interacting safely with humans, is at present more imperative than ever.

This book presents a collection of ten chapters that address some of the latest advances in HRC, with a focus on achieving efficient, safe and genuine HRC. The chapters explore industry requirements and identify the necessary steps to improve HRC research and its adoption by industry in several industrial application domains.

Chapter 1 provides a brief introduction to HRC, discussing the limitations of traditional industrial robots and highlighting some of the most important advances in HRC research. Chapters 2 through 10 present the state-of-the-art research result in HRC. In their majority, these peer-reviewed chapters are the outcome of the workshop "Unlocking the potential of HRC in industrial applications," which was organized by the editors during the International Conference on Robotics and Automation (ICRA) in 2021. Five of the chapters present novel research results conducted within the CoLLaboratE project funded by the European Union's Horizon 2020 programme. The book concludes with a wrap up and future work epilogue by the editors, where open research questions and future research directions are presented.

This book can be a helpful resource for researchers, practitioners and students in the field of robotics and automation, providing insights into the latest advances in HRC and the potential it offers for industrial applications.

Chapter 1

Introduction to human-robot collaboration for industrial applications

Fotios Dimeas[1] *and Zoe Doulgeri*[1]

Human-robot collaboration (HRC) is a widely studied research field that has gained attention during the past few years with the ever-increasing availability of collaborative robots (or cobots). Cobots have the sensing capabilities to actively interact with a human partner toward achieving a common task. Some forms of interaction involve direct physical contact, often referred to as physical human-robot interaction (pHRI). A strong focus of the pHRI-related literature is on ensuring safety during the interaction. Despite the continuous advances in HRC research, applications are few and are usually downgraded to basic workspace sharing rather than actual collaborative work. Taking full advantage of cobots in tasks such as collaborative manipulation of heavy objects or collaborative assembly operations has yet to demonstrate tangible results for the industry, as there is still a need for cobots to operate in an intelligent and adaptive way and be easily programmed by nonexperts. This book includes a series of chapters, presenting some of the latest advances in HRC in an effort to unlock its potential for industrial adoption. The book investigates what the needs of the industry are and what steps should be taken in the research toward efficient, safe, and useful HRC.

1.1 Background

The majority of the robots that are currently employed in the industry operate autonomously behind fences in well-structured environments, following a preplanned trajectory to achieve a task. They are rigid body constructions with limited sensing abilities, moving at high speed with precision and repeatability. These robots are potentially dangerous for humans, and for this reason, they are separated from them. Since the 1970s, such robots have been continuously replacing repetitive, time-intensive, or even menial tasks that were previously done manually, with the aim of reducing production costs, increasing production volume, and improving working conditions.

[1]Automation and Robotics Lab, Aristotle University of Thessaloniki, Thessaloniki, Greece

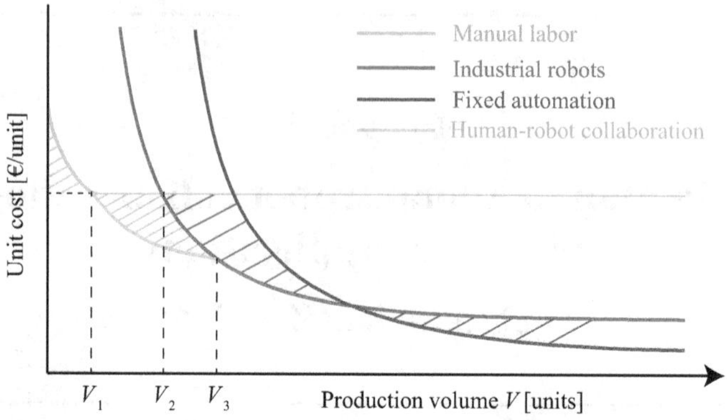

Figure 1.1 *Chart of unit cost versus production volume [3]*

Typical applications of robots are free-space movements with almost no interaction with the environment (painting, welding, and quality control) and pick-and-place movements with some interaction but in well-structured environments (palletizing, sorting, and machine tending). More recently, robots have been employed in assembly operations that involve the development of significant interaction forces and require appropriate robot reactions in order to carry out the task [1, 2].

Industrial robots, although aimed at flexible production systems, have a limited perception of their environment compared to human capabilities and require high reprogramming times. Consequently, the installation and operation of a robot are justified only when there is a significant volume of production, as shown in Figure 1.1. At a very small production volume (up to V_1), manual work is chosen because the cost of automation is prohibitive. This is due to the fact that humans are able to quickly learn complex tasks, are better at understanding changes in the environment or the task, and have rapid adaptation abilities. However, unlike robots, humans have lower strength and speed, and low repeatability, while the fatigue factor has a negative effect on the result of a consistent quality. Robot usage is justified for significantly higher production volumes, above V_2. As a result, there is a flat area between V_1 and V_2, where the unit cost of manual labor has reached a lower bound, although production volume has increased. This is where HRC comes into play.

The term *human-robot cooperation* is a special category of the broader term *human-robot interaction*, which includes the concept of the joint action of two or more agents to achieve a goal. An additional term found in the literature is *human-robot collaboration*, which specifies that the purpose of the interaction is production. In the ISO 10218:2011 [4] standardization aimed at industrial applications, human-robot collaborative operation is defined as *the situation in which specially designed robots work in direct cooperation with humans within a defined workplace*. The revised technical specification ISO/TS 15066:2016 [5] for human-robot collaboration lists safety measures that should be considered as well as challenges that should be addressed for robots to work alongside and/or in collaboration with humans.

Human and robot strengths are complementary, and their synergy can bridge the industrial gap between manual labor and robotic automation even in cases of medium production volumes (between V_2 and V_3 in Figure 1.1). In particular, a robot working together with a human can assist the human by undertaking tasks that can be repetitive or require effort or precision levels that exceed those of a human. When there is physical contact between them, the robot can reduce the human's physical effort by load sharing, when, for example, large and heavy object handling is required, as well as enhance precision and ease of handling via its partial autonomy. Moreover, it can be easily programmed by demonstration, thus learning to perform variants of the learned task via spatial and temporal scaling. For example, a skilled worker can kinesthetically demonstrate a task to the robot by grabbing it and showing it how to move, thus encoding a richer pattern of motions than those achieved by conventional programming, and dramatically reducing programming time.

1.2 Unlocking the potential

HRC is an important element of Industry 4.0, the industry trend that is also fueled by advances in artificial intelligence (AI), additive manufacturing, and the Internet of Things. With the more human-centric Industry 5.0 underway, which focuses on cooperation between humans and machines, collaborative robots will play an even more significant role. This is testified by the literature on HRC over the last two decades, which has received an important boost in the past 10 years, as shown in the number of publications on this topic in Figure 1.2. Moreover, most of the robot manufacturers have included collaborative manipulators in their catalouge that are provided with basic state-of-the-art sensing and safety functionalities. However, to unlock the full potential of industrial HRC, recent hardware, sensing, and control advances need to be integrated into these robots to make them capable of adapting

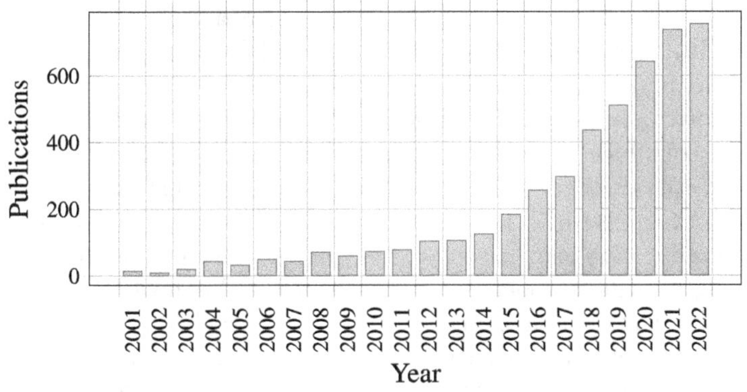

Figure 1.2 *The number of publications on the topic of HRC from 2001 to 2022. Data were obtained from Scopus using the search (human AND robot AND collaboration) in title, abstract, and keywords.*

their autonomy level based on their own understanding of human actions and intentions and of their dynamically changing environment.

Starting from the hardware level, new actuators have come into play that allow us to move away from the traditional stiff position-controlled paradigm in which robots are ignorant of external contact forces, into torque-controlled robots with force/torque sensing capabilities that can measure and adapt to external forces applied by humans or the environment. Actuators with back-drivability and variable stiffness, coupled with new control designs, can enhance safety and performance in human-robot collaborative tasks. However, to be truly collaborative, the robot must have the ability to understand the human co-worker. To achieve this aim, the use of vision and proximity sensors can be considered as the first stage of interaction modality needed in order for the robot to detect human presence in its vicinity. Detection of gestures through vision can be an example of high-level cooperation for the robot to react to human commands. For closer interaction and true physical HRC, the main robot perception modality is the use of force, tactile sensors, and even robot skins that cover important parts of the robot body in order for the robot to detect human contact and adapt its behavior accordingly by regulating the cooperation force. Such modalities have been extensively studied in the literature and partially adopted by the industry to provide collaborative robots with enhanced capabilities [6]. Other modalities such as electromyography, electroencephalography, or a combination of all the aforementioned modalities have been proposed but, even fewer applications can be found in the industry due to their invasive nature. In the opposite direction, various modalities have been developed in the literature to help human co-workers understand the actions of the robot, such as haptic feedback or visual feedback approaches that are enhanced with augmented reality.

Shared control allows a human operator and an autonomous controller to simultaneously and collaboratively control the robotic system to achieve a common goal, combining the robot's precision with the superior scene understanding and cognitive capabilities of humans. The robot's autonomy level can be either predefined depending on the task or automatically adjusted based on the robot's understanding of human intentions, its knowledge of the task, and the surrounding environments [7]. The latter is also termed as shared autonomy.

At the lower level of predefined autonomy, the human is fully responsible for controlling the tool motion by applying forces and torques to the robot holding the tool either directly (hands-on) or via a haptic device in a teleoperation set-up. The objective of the robot's controller is, in this case, to satisfy human intent by preventing motion into predefined regions and creating a stable natural motion closer to the feeling of manipulating a weightless tool unattached to the robot. This low level of autonomy is the only option in many unstructured and safety-critical applications where any failure cannot be tolerated and highly dexterous movements are required, for example, nuclear scenarios, robotic surgery, or manipulation of large and heavy objects for inspection and assembly. It is also utilized during the teaching phase of learning a task by kinesthetic demonstration to provide haptic signals to the user and thus increase his/her awareness about the proximity to constraints or the current learning level [8].

Automatically adapting the autonomy level requires correctly inferring the human's intentions and task progress, which raises new challenges and requires novel control structures. An example of this application is the progressive automation of discrete and periodic tasks, which unifies the teaching, learning, and execution phases within the context of learning by demonstration [9, 10]. The basic core of shared control architectures consists of impedance and/or admittance control modalities, which are the standard way to regulate the robot's behavior under external force inputs [11]. Their combination with recent developments in sensing, inference, and learning methods has allowed the development of effective shared control strategies that can widen the spectrum of HRC applications.

Despite the promising technologies being developed, there can be issues that slow down their adoption due to liability, certification, standardization, and human trust. A robotic system that consists of multiple components (arm, controller, sensors, AI algorithms, and middleware) with the ability to extend its actions in new environments has uncertain liability in case of an accident. Certification is a related impediment because standards are sparsely updated and not mature enough, restricting fenceless collaboration. Certification is also required and given for a certain application rather than a range of applications that can be covered by a collaborative robot. Last, in a robotic system incorporating AI for decision-making, the lack of formal proof of their behavior and transparency and explanation of their reasoning is a further impediment to certification. Human trust is another essential aspect of the adoption of HRC. Trust toward a cobot depends on its features and performance in a particular application, as well as the human's attitude and experience.

Nevertheless, national and cross-national research projects are being funded for the development of HRC technologies, consisting of consortia with academic researchers, technology providers, robotic integrators, and industrial end-users. Through these projects, end-users provide real-world requirements and set up demonstrations targeting real-world applications and scenarios to validate the technologies beyond laboratory setups and test their robustness and reliability. These efforts increase the readiness levels of human-robot collaborative technologies and pave their way to the market.

1.3 Book objectives

In this book, we will investigate some of the latest advances in HRC with industrial applications, aiming to move a step further in unlocking their potential. The book consists of a collection of nine peer-reviewed chapters by academics and researchers. It is the follow-up of the workshop "Unlocking the potential of human-robot collaboration in industrial applications." This workshop was organized by the editors during the International Conference on Robotics & Automation (ICRA) 2021. Selected speakers and short-paper contributors were invited to participate in this book with an extended version of their novel research.

Teaching by demonstration is a technique that has many applications since robots acquire manipulation skills through interaction with humans. *Sylvain Calinon* discusses the challenges of such techniques in the Chapter 2. He presents intuitive interfaces to acquire meaningful demonstrations, methods for efficient representations of manipulation skills, and ways to adapt the learned skills to new situations, task variations, or perception uncertainties. This chapter concludes with the barriers that still need to be overcome to move these techniques out of the laboratory and into the real world.

Taking into consideration a real-world industrial scenario of a TV assembly, a control scheme is presented by *Leonidas Koutras et al.* in the Chapter 3. This method presents new formulations of Dynamic Movement Primitives and discusses how they can enable adaptive execution by appropriate scaling of the demonstrated trajectory to new and even moving targets and addressing the problem of assembly lines where operations take place on a moving conveyor belt. The chapter also addresses the safety aspect, suggesting two layers for human safety preservation: the first in the form of collision avoidance by altering the produced trajectory, and the second with the compliant behavior of the arm, but without compromising reaching accuracy.

On the same testbed as the TV assembly, *Georgia Peleka et al.* dive into the perception aspect of safe HRC in the Chapter 4. They discuss the problem of how to detect humans when collaborating closely with the robot, even under occlusions, how to model the humans, and how to track them in 3D for their subsequent avoidance by the robot. The safety aspect is also discussed by *Alberto Olivares-Alarcos et al.* in the Chapter 5. Having in mind the separation distance proposed by the ISO standards, this chapter tries to address the problem of close HRC by means of velocity adaptation and a safety-rated monitored stop, without altering the nominal trajectory of the robot.

Motion coordination between a human and a robot is necessary when both are performing a task that requires true collaboration. On this aspect, *Andrea Pupa et al.* discuss task scheduling and safety in the Chapter 6. The authors present a framework so that the robot is aware of human activities during collaboration, dynamically reschedules the task execution strategy, and adjusts the robot's trajectory toward safe and natural collaboration. Also with the aim of coordination, methods to synthesize new movements from a set of example demonstrated trajectories that are presented by *Miha Deniša* and *Aleš Ude* in the Chapter 8. With an approach based on a hierarchical database, a robot can recognize the movement of a human and perform the demonstrated task in collaboration with them in a coordinated and smooth manner.

On the hardware aspect of HRC, a tactile skin is presented by *Francesco Grella et al.* in the Chapter 7. The authors discuss how this novel skin can be used as an enabler for collaborating with heavy industrial robots and how it can be combined with AI for a robust distinction between voluntary and involuntary contacts of the human with the robot. The chapter also presents the integration of the skin in a real-world task with a heavy industrial robot.

An application-oriented system that leverages teaching by demonstration is presented by *Antonakoudis et al.* in the Chapter 9. Using teaching by demonstration, the user can create the robot's pick-and-place motion pattern in a very short time with zero-code programming. Machine vision with AI provides the robot with the ability to recognize its environment, adapt its movements, and remain safe for human interaction. This versatile robotic system can be easily and quickly taught by personnel how to pack products on a production line.

In the final Chapter 10 by *Yunus Terzioğlu et al.*, the authors provide some of their latest advances on HRC. Focusing on assembly tasks, the chapter presents methods for how the robot can become a useful assistant to the human for handing over tools. Special attention is given to the social aspects and robot acceptance, considering different modalities such as human pose and gaze estimations for more efficient and human-friendly interaction.

Five of the chapters present novel research results that were conducted within the CoLLaboratE* project, which was funded by the European Union's Horizon 2020 program. This project aimed to revolutionize the way industrial robots learn to cooperate with human workers when performing assembly tasks. New methods were developed to equip robots with collaborative skills, using movement primitives, deep reinforcement learning algorithms, and safety strategies, in combination with new safety skin hardware. The consortium, including universities, research institutes, small and medium-size enterprises, and industries, managed to integrate and validate the results of the projects in four real-world use cases from the home appliance, automotive, and aircraft industries.

References

[1] Zhu Z., Hu H. 'Robot learning from demonstration in robotic assembly: a survey'. *Robotics*. 2018, vol. 7(2), p. 17.

[2] Faccio M., Bottin M., Rosati G. 'Collaborative and traditional robotic assembly: a comparison model'. *The International Journal of Advanced Manufacturing Technology*. 2019, vol. 102(5–8), pp. 1355–72.

[3] Bjorn M. 'Industrial safety requirements for collaborative robots and applications. ERF 2014 – workshop: Workspace safety in industrial robotics: trends, integration and standards' in ABB Corporate Research; 2014.

[4] ISO. ISO 10218-1 Robots and robotic devices – safety requirements for industrial robots – Part 1: Robots. The International Organization for Standardization; 2011.

[5] ISO. ISO/TS 15066 Robots and robotic devices – collaborative robots. The International Organization for Standardization; 2016.

[6] Ajoudani A., Zanchettin A.M., Ivaldi S., Albu-Schäffer A., Kosuge K., Khatib O. 'Progress and prospects of the human-robot collaboration'. *Autonomous Robots*. 2018, vol. 42(5), pp. 957–75.

[7] Selvaggio M., Cognetti M., Nikolaidis S., Ivaldi S., Siciliano B. 'Autonomy in physical human-robot interaction: a brief survey'. *IEEE Robotics and Automation Letters*. 2021, vol. 6(4), pp. 7989–96.

*https://collaborate-project.eu

[8] Papageorgiou D., Kastritsi T., Doulgeri Z. 'A passive robot controller aiding human coaching for kinematic behavior modifications'. *Robotics and Computer-Integrated Manufacturing*. 2020, vol. 61, 101824.
[9] Kastritsi T., Dimeas F., Doulgeri Z. 'Progressive automation with DMP synchronization and variable stiffness control'. *IEEE Robotics and Automation Letters*. 2018, vol. 3(4), pp. 3789–96.
[10] Dimeas F., Doulgeri Z. 'Progressive automation of periodic tasks on planar surfaces of unknown pose with hybrid force/position control'. *2020 IEEE/RSJ International Conference on Intelligent Robots and Systems (IROS)*; Las Vegas, NV, USA, IEEE, 2020. pp. 5246–52.
[11] Keemink A.Q., van der Kooij H., Stienen A.H. 'Admittance control for physical human-robot interaction'. *The International Journal of Robotics Research*. 2018, vol. 37(11), pp. 1421–44.

Chapter 2

Programming industrial robots from few demonstrations

Sylvain Calinon[1]

2.1 Introduction

A wide range of industrial applications can benefit from robots acquiring manipulation skills by interaction with humans. This chapter discusses the challenges that such a learning process encompasses, including the development of intuitive interfaces to acquire meaningful demonstrations, the development of representations for manipulation skills exploiting the structure and geometry of the acquired data in an efficient way, and the development of optimization and optimal control techniques that can adapt to new situations and that can exploit coordination, task variations, and perception uncertainties. The chapter presents an overview of the research lines of the *Robot Learning and Interaction* group at the *Idiap Research Institute*. It analyses the barriers that still need to be overcome to move these techniques out of the laboratories by complying with real industrial application constraints. It also proposes a variety of research directions for which further investigations would be required to overcome these barriers.

Industrial robotics can leverage the development of robots that can acquire new skills from only a few demonstrations and interactions. This challenge is multifaceted and can be addressed from varied perspectives. It often requires a wide range of research fields to be considered, as diverse as human-robot collaboration, interface design, movement primitives (MP), motor skills, imitation learning, shared control, optimization, and planning. This chapter focuses on the *learning from demonstration* (LfD) aspects [1]. It takes a practical stance by narrowing down the range of possible approaches by targeting the use of only a *few* human demonstrations to program a robot so that it can acquire new manipulation tasks fast and adapt the learned knowledge to changing situations.

[1]Idiap Research Institute, Martigny, Switzerland

2.2 Learning in a handful of trials

The field of machine learning has evolved toward approaches relying on increasingly large datasets. In several application domains, these datasets are already available or are relatively inexpensive to collect/produce, which motivated such an evolution. Robotics in industrial applications are characterized by a different problem setting, which contrasts to this race of using bigger and bigger amount of data in machine learning. It can be viewed as a *small data learning problem*, or as a wide-ranging learning data problem [2], with models that can start learning from small datasets and that can still exploit additional data if such data becomes available during the robot's lifespan.

In contrast to other fields, the data formats in robotics vary significantly between tasks, environments, users, and platforms (different sensors and actuators, not only in formats but also in modalities and organizations). The learned models often need to be interpretable to provide guarantees and to be linked to other techniques. This chapter motivates that research in industrial robot applications should target the development of learning approaches that can rely on only few demonstrations or trials.

To achieve this research agenda, the main challenges to target boil down to finding structures that can be used in a wide range of tasks, which include (from high level to low level):

- Learning structures and interfaces (§2.3)
- Combination structures (§2.4)
- Geometric structures (§2.5)
- Data structures (§2.6)

which are described in the next sections of this chapter (see also Reference 3 for a survey concentrating on the learning aspects). Examples will be used throughout the chapter to show that efficient manipulation and assembly skills acquisition require a suitable trade-off between learning and exploitation of these different forms of structures (at model and algorithm levels).

Section 2.3 highlights the importance of considering bidirectional interaction in the learning process and the importance of providing user-friendly interfaces to facilitate learning. An example is given with a smartphone interface using augmented reality (AR) for a collaborative riveting task. Section 2.4 presents extensions of MP representations. It includes task-parameterized models, which are object-centric representations of movements, and a probabilistic optimal control formulation of dynamical MP (DMP). An example is given for the teleoperation of a distant robot opening valves. Section 2.5 discusses the importance of geometric representations. An example is given for the transfer of manipulation skill in a screwing task by leveraging the use of manipulability ellipsoids as geometric descriptors. Section 2.6 finally highlights the roles of tensor factorization to extract essential information from high-dimensional data. An example is given with a peg-in-hole insertion task learned from the demonstration by leveraging the use of ergodic control to efficiently search for the insertion point.

2.3 Learning structures and interfaces

To reduce the amount of required data, a first opportunity to seize is that machine learning in robotics goes beyond the standard training- and testing-set paradigm. Skills acquisition in robotics is a *scaffolding process* rather than a bland learning process. The robot needs a lot of structures at the beginning, and the structures can be progressively dismantled when the robot progresses: there is a continuous evolution from full assistance to full autonomy. We can thus exploit a number of interactive learning mechanisms to acquire/generate better data on the spot, including active learning, machine teaching (by generating data to train the robots), curriculum learning (by providing data of increased complexity adapted to the robot learner), and bilateral interactions that rely on several social mechanisms to transfer skills more efficiently, see Figure 2.1.

To acquire manipulation skills, humans greatly benefit from the combination of several learning modalities. Similarly, multiple learning strategies can jointly be exploited to transfer skills to industrial robots while facilitating the evaluation of the robot's current capability at executing the task by the operator [4, 5]. The research direction of orchestrating different learning modalities is still in its infancy in robotics research. Currently, it essentially targets experiments in research laboratories. However, for industrial robot applications, this research direction has the potential to improve the robustness of the acquired manipulation and assembly skills by

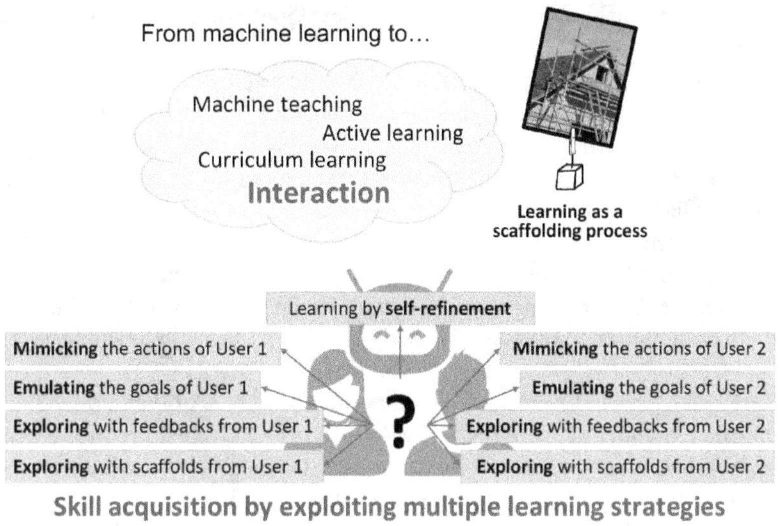

Figure 2.1 An orchestration of varied learning strategies is required to efficiently acquire new skills

allowing diverse forms of constraints to be considered (at the level of the users and the environments), as well as improved monitoring of the robot knowledge (by testing the acquired skills in diverse situations).

The other important advantage of combining learning strategies is that it allows each individual learning modality to be simplified. Indeed, skills acquisition with a single learning strategy can be unnecessarily complex. As an illustration, we cannot learn to play a sport efficiently without practice by only observing others play. Similarly, in a manufacturing environment, we cannot acquire efficiently a fabrication skill by limiting the available information to only end goals (or final products). We instead need to observe an expert or to be guided throughout the process to acquire the underlying fabrication strategies. Similarly to us, robots might not be able to acquire skills efficiently by using a single learning modality. Thus, instead of focusing on the improvement of algorithms and models for a specific learning strategy, industrial robotics could highly benefit from further research to investigate the meta-learning problem of combining efficiently existing learning modalities, without defining the sequence and/or organization in advance [6].

2.3.1 User interfaces for skill transfer

Robots acquiring skills by user guidance require the development of user-friendly interfaces for the operators to teach and monitor the acquisition of manipulation skills. First, this is important to let the operators provide useful demonstrations of the task by allowing them to concentrate on the skill to transfer instead of nonessential hindering factors. This is also important to let them assess the current capability of the robot in order to provide demonstrations or corrections that are the most appropriate with respect to the current capability of the robot [4, 5]. Thus, interfaces enabling bidirectional interactions need to be developed within the context of industrial applications.

Currently, the most commonly used interfaces to teach robots new tasks in industrial applications are the *teaching pendants*. Recent torque-controlled or collaborative robot manipulators brought a new programming interaction capability by letting operators drive the robots manually, with backdrivable motors and controllers that compensate for the effect of gravity. The resulting interfaces allow users to move robots freely in space as if there was no motorization in the articulations and as if the robots had no weight. This *kinesthetic teaching* capability can be extended to more elaborated forms of virtual fixtures by exploiting impedance through the torque control capability of the robot, which provides a physical guidance capability to facilitate manipulation skill transfer [7].

Research in *virtual reality* (VR), *augmented reality* (AR), and *mixed reality* (MR), gathered under the name of *extended reality* (XR), offers promising alternative approaches to interact with robots for both teaching and monitoring aspects. A subset of these approaches rely on the AR capability of recent smartphones or tablets by displaying the video stream from the cameras on the back of the devices, and by superposing to these images virtual robots and/or virtual guides (typically, paths, labels, or coordinate systems), see Reference 8 for a review.

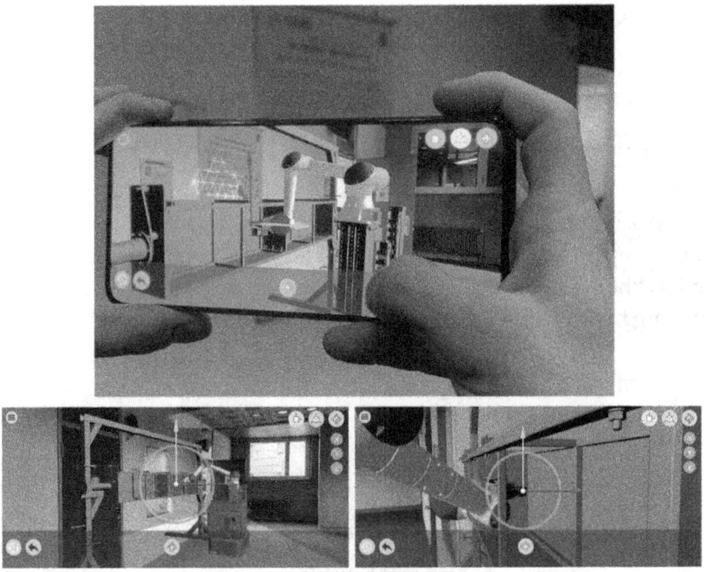

Figure 2.2 *Smartphone interface using AR capability, exploited in an industrial application requiring the operator to program a series of riveting poses to a six-axis HC10 Yaskawa manipulator fixed to an elevating platform mounted on an Automated Guided Vehicle (AGV)*

The smartphone/tablet looks like a transparent window that the operator can freely move to observe a virtual scene superposed to the real scene. By changing the angle of view and by moving closer/further away, the user can intuitively analyze existing movements and program new robot movements. It facilitates the inspection of a taught path by replaying consecutively a robot motion on the virtual robot before executing it on the real robot.

In Reference 9, this approach was tested in the context of industrial (de)insertion tasks with a NIST Task Board [10], a benchmark platform used to evaluate assembly performance. Figure 2.2 presents the use of this AR interface within an industrial application. This case study is part of the CoLLaboratE European project[*] targeting the use of cobots in industrial applications. This case study involves the float of an aircraft to be riveted in collaboration with the operator. The compartments of the aircraft float require a careful plan of the robot movements in order to reach robot configurations that are convenient for collaborative riveting and that do not collide with the aircraft structure during the motion between any two consecutive rivets. Here, the developed AR interface is used to let the operator program and monitor the robot motion plans, either from a remote location

[*]https://collaborate-project.eu/

(i.e., without access to the robot and aircraft float, as depicted in Figure 2.2) or onsite the industry without the robot (i.e., virtual robot superposed to the real aircraft float).

2.4 Combination structures

The term *movement primitives* refers to an organization of continuous movements in the form of a superposition in parallel and in a series of simpler signals, which can be viewed as "building blocks" to create more complex movements. This principle of superposing high-level "bricks of motion" from a dictionary, coined in the context of motor control [11], remains valid for a wide range of continuous-time signals (for both analysis and synthesis). In particular, this notion can be extended to *behavior primitives*, which form a richer set of assistive behaviors. These behavior primitives correspond to controllers that are either myopic or anticipative, with either time-independent or time-dependent formulations, see Figure 2.3 for an overview.

The combination of behavior primitives can be formalized as an information fusion problem in which several sources of information can be modeled as probability distributions [12]. This results in a *product of experts* (PoE) modeling a probability distribution by combining the output from several simpler distributions [13]. The core idea is to combine several distributions (called experts) by multiplying their density functions. This allows each expert to make decisions on the basis of a few dimensions, without having to cover the full dimensionality of a problem. Illustratively, a PoE corresponds to an "and" operation, which contrasts with a mixture model that corresponds to an "or" operation (by combining several probability distributions as a weighted sum of their density functions). Thus, each component in a PoE represents a soft constraint. For an event to be likely under a product model, all constraints must be (approximately) satisfied. In contrast, in a mixture model, an event is likely if it matches (approximately) with any single expert.

With Gaussian distributions, the fusion problem simplifies to a *product of Gaussians* (PoG), which can be solved analytically, where the distributions can either represent robot commands at the current time step (myopic control system) or trajectory distributions in the control space (anticipative planning system) [14].

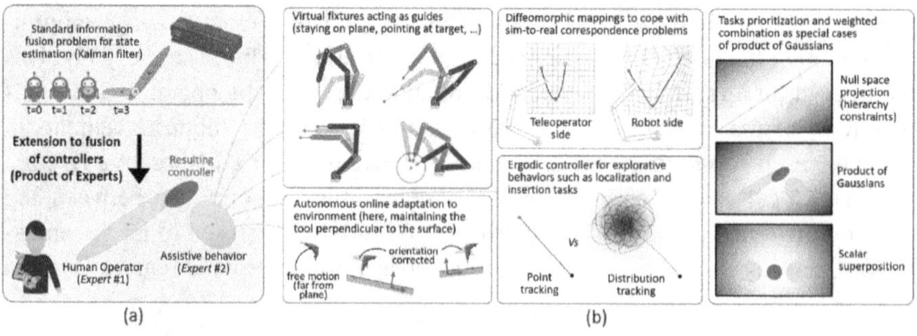

Figure 2.3 Examples of combination structures (see main text for details)

State estimation is an example of problems classically solved as an information fusion problem, resulting in a PoG that takes into account uncertainty in motion and sensor(s) models (a well-known example is the Kalman filter), see Figure 2.3(a). A control problem in a collaborative industrial task can be treated in a similar mathematical framework at the control level. This approach allows multiple behavior primitives to be combined in parallel (including prioritization with nullspace projection structures, see Reference 12 for details). In this framework, each operator (human or artificial) contributes to the task by taking into account the (co)variations allowed in a task and the uncertainty of the expertise. In essence, this shared control strategy has links with products of Gaussians and Riemannian motion approaches, but the PoE provides a more general formalism (including non-Gaussian distributions) and allows the same strategy to be considered for both shared perception and shared control paradigms.

For industrial applications, this approach allows the combination of different controllers, which can be learned separately or altogether (by variational inference, see Reference 12 for details). With this formulation, a robot can counteract perturbations that have an impact on the fulfillment of the task while ignoring other perturbations. This approach relates to research in biomechanics and motor control, with formulations including the principle of synergies [15], *minimal intervention principles* [16], *uncontrolled manifolds* [17], or *optimal feedback control* [18]. This formulation also opens the road to a rich set of challenges that can be investigated in manipulation and assembly with either autonomous robots or cobots, including bimanual manipulation skills, the handling of both force and position information, as well as the simultaneous handling of multiple objectives (including physical compliance, manipulability, safety, and task hierarchies). Figure 2.3(b) presents examples of assistive behaviors in the context of manipulation and assembly skills.

Next, the PoE/PoG principle is illustrated with the problem of fusing information from multiple coordinate systems.

2.4.1 *Task-parameterized movements*

To facilitate the acquisition of manipulation skills, *task-parameterized models* can be exploited to take into account that motions typically relate to objects, tools, or landmarks in the robot's workspace, see Reference 19 for a review and Reference 20 for an industrial case study.

The approach consists of encoding a movement in multiple coordinate systems (namely, from the perspectives of different objects) in the form of trajectory distributions. In a new situation (i.e., for new object locations), the reproduction problem corresponds to a fusion problem, where the variations in the different coordinate systems are exploited to generate a movement reference tracked with variable gains, providing the robot with a variable impedance behavior that automatically adapts to the precision required in the different phases of the task. For example, in a pick-and-place task, the robot will be stiff if the object needs to be reached/dropped in a precise way and will remain compliant in the other parts of the task.

For user assistance in industrial operations, this approach can be used for local adaptation to objects/landmarks in the robot's environment, such as automatically correcting the orientation of a tool held by the operator to ensure that it remains perpendicular to a surface, see Figure 2.3(b).

Figure 2.4 presents an example of an application for remote valve turning operations, developed in the context of the DexROV European project [22]. In this example, a *task-parameterized Gaussian mixture model* (TP-GMM) approach is used within an optimal control strategy to teleoperate the arms of a bimanual underwater robot from a distance, with a teleoperator wearing an exoskeleton and visualizing a copy of the robot workspace in a virtual environment. Because of the long communication delays between the teleoperator and the robot, the locations of the objects or tools of interest are not the same on the teleoperator side and on the robot side. With the proposed parameterization that leverages the locations of objects and tools, this discrepancy can be handled by automatic adaptation of the movement representation to the position and orientation of the objects/tools, represented as coordinate systems.

The central part of Figure 2.4 depicts two coordinate systems. The two components in red correspond to the robot and to the valve to be turned for both position and orientation data. The encoding of the movements with respect to the robot and to the valve is depicted in orange and in purple, respectively.

The movements relative to the valve and to the robot are encoded as two sets of Gaussians in the two coordinate systems for both position and orientation. During teleoperation, the distributions are rotated and translated according to the current

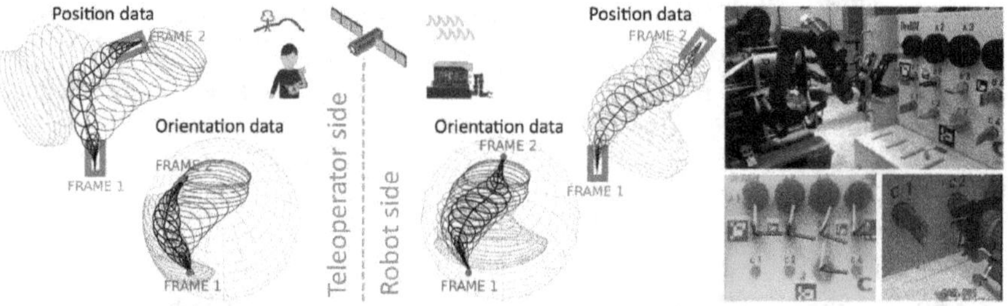

Figure 2.4 *Task-parameterized movement models used within an offshore oil field application involving valves to be reached and turned remotely by a teleoperator. The purple and orange ellipsoids depict two GMMs, representing uncertain trajectories with respect to two different frames of reference (red U shapes for position data and red points on the spheres for orientation data). The black ellipsoids represent the final trajectory and its uncertainty, obtained by fusing the trajectories of the two different frames of reference through products of Gaussians. Although the red U shapes and red points are not the same on the teleoperator side and on the robot side, the retrieved paths on the two sides can very quickly adapt to these different situations (see main text for details). Figure adapted from Reference 21.*

situation on the teleoperator side and on the robot side. Products of Gaussians are computed at each side to fuse these representations (depicted in black in the figure).

Movement is encoded in this way for both position $\mathbb{R}^3$ and orientation $\mathcal{S}^3$ data (in Figure 2.4, a representation with $\mathbb{R}^2$ and $\mathcal{S}^2$ is shown as an illustration). For position, the retrieved path in black (and the associated covariances) corresponds to a movement going from the robot to the valve (represented as red U shapes) by taking into account how these different coordinate systems are oriented. We can see that the approaching phase is perpendicular to the coordinate system to properly reach the valve. For orientation, the retrieved path shows how the orientation of the end effector changes with time. At the beginning of the motion, this orientation relates to the orientation of the robot, while at the end, the orientation of the end effector matches the orientation of the valve.

This task-parameterized modeling approach was successfully tested in field trials in the Mediterranean Sea offshore of Marseille, where seven extended dives in four different sites (8, 30, 48, and 100 m water depths) were performed with the underwater robot while being connected via satellite to the teleoperation center in Brussels, see Reference 22 for an overview of the experiment.

The task-parameterized principle can be used as a first starting point to study the fusion of more elaborated coordinate systems. In Figure 2.4, simple coordinate systems centered on several points of interest were considered, but the approach could be extended to a richer set of behaviors, including a wider range of coordinate systems that take into account symmetries such as cylindrical and spherical coordinate systems [23] and nullspace projection structures [24].

In the next section, I will show that the description of movements as trajectory distributions can be extended to control problems by adopting an optimal control formulation with costs expressed as simple quadratic error terms.

2.4.2 Probabilistic optimal control

Skill acquisition encompasses a broad spectrum of learning approaches, from action-level *imitation* to goal-level *emulation*, where the mutual benefits of these two learning strategies have been highlighted in animals and humans studies [25].

While *emulation* requires higher cognitive skills to infer the intended goals behind a set of actions, *imitation* allows to acquire skills in a simple but limited way by directly mimicking the movements and actions shown to the learner. Because of its higher-level nature, *emulation* can easily be misinterpreted as being the only useful mechanism. Studies such as Reference 25 show that the two are complementary, in the sense that *emulation* allows greater generalization (i.e., the learner understands the purpose of the task and can reproduce it in possibly different ways), while imitation allows to acquire tasks fast, even if all underlying goals are not fully understood (*"copy all, refine later"* strategy).

Similarly, in the engineering world, research in optimal control and reinforcement learning is tightly connected to the *emulation* approach, while research in MP is tightly connected to the *imitation* approach. The mutual benefit can be illustrated by emulation making the system more generalizable, while imitation/mimicking can improve the learning rate, see References 5, 26, and 27.

I will show next that optimal control provides a formulation based on a cost that implements the goal-level nature of *emulation*, while control primitives (CP) provide a formulation based on basis functions that implement the action-level nature of *imitation*.

2.4.2.1 Linear quadratic tracking

Linear quadratic tracking (LQT) and *linear quadratic regulation* (LQR) are the simplest forms of optimal control that trade-off tracking and control costs expressed as quadratic terms over a time horizon, with the evolution of the state described in a linear form. The LQT problem is formulated as the minimization of the cost:

$$\begin{aligned} c &= \left(\boldsymbol{\mu}_T - \boldsymbol{x}_T\right)^\top \boldsymbol{Q}_T \left(\boldsymbol{\mu}_T - \boldsymbol{x}_T\right) + \sum_{t=1}^{T-1} \left(\boldsymbol{\mu}_t - \boldsymbol{x}_t\right)^\top \boldsymbol{Q}_t \left(\boldsymbol{\mu}_t - \boldsymbol{x}_t\right) + \boldsymbol{u}_t^\top \boldsymbol{R}_t \, \boldsymbol{u}_t \\ &= \left(\boldsymbol{\mu} - \boldsymbol{x}\right)^\top \boldsymbol{Q} \left(\boldsymbol{\mu} - \boldsymbol{x}\right) + \boldsymbol{u}^\top \boldsymbol{R} \boldsymbol{u} \end{aligned} \tag{2.1}$$

with $\boldsymbol{x} = \left[\boldsymbol{x}_1^\top, \boldsymbol{x}_2^\top, \ldots, \boldsymbol{x}_T^\top\right]^\top$ the evolution of the state variables, $\boldsymbol{u} = \left[\boldsymbol{u}_1^\top, \boldsymbol{u}_2^\top, \ldots, \boldsymbol{u}_{T-1}^\top\right]^\top$ the evolution of the control commands, and $\boldsymbol{\mu} = \left[\boldsymbol{\mu}_1^\top, \boldsymbol{\mu}_2^\top, \ldots, \boldsymbol{\mu}_T^\top\right]^\top$ the evolution of the tracking targets. $\boldsymbol{Q}=\text{blockdiag}(\boldsymbol{Q}_1, \boldsymbol{Q}_2, \ldots, \boldsymbol{Q}_T)$ represents the evolution of the precision matrices $\boldsymbol{Q}_t$, and $\boldsymbol{R}=\text{blockdiag}(\boldsymbol{R}_1, \boldsymbol{R}_2, \ldots, \boldsymbol{R}_{T-1})$ represents the evolution of the control weight matrices $\boldsymbol{R}_t$.

The evolution of the system is linear, described by $\boldsymbol{x}_{t+1} = \boldsymbol{A}_t \boldsymbol{x}_t + \boldsymbol{B}_t \boldsymbol{u}_t$, yielding $\boldsymbol{x} = \boldsymbol{S}_{\boldsymbol{x}} \boldsymbol{x}_1 + \boldsymbol{S}_{\boldsymbol{u}} \boldsymbol{u}$ at trajectory level, with

$$\underbrace{\begin{bmatrix} \boldsymbol{x}_1 \\ \boldsymbol{x}_2 \\ \boldsymbol{x}_3 \\ \vdots \\ \boldsymbol{x}_T \end{bmatrix}}_{\boldsymbol{x}} = \underbrace{\begin{bmatrix} \boldsymbol{I} \\ \boldsymbol{A}_1 \\ \boldsymbol{A}_2 \boldsymbol{A}_1 \\ \vdots \\ \prod_{t=1}^{T-1} \boldsymbol{A}_{T-t} \end{bmatrix}}_{\boldsymbol{S}_{\boldsymbol{x}}} \boldsymbol{x}_1 + \underbrace{\begin{bmatrix} \boldsymbol{0} & \boldsymbol{0} & \cdots & \boldsymbol{0} \\ \boldsymbol{B}_1 & \boldsymbol{0} & \cdots & \boldsymbol{0} \\ \boldsymbol{A}_2 \boldsymbol{B}_1 & \boldsymbol{B}_2 & \cdots & \boldsymbol{0} \\ \vdots & \vdots & \ddots & \vdots \\ \left(\prod_{t=1}^{T-2} \boldsymbol{A}_{T-t}\right) \boldsymbol{B}_1 & \left(\prod_{t=1}^{T-3} \boldsymbol{A}_{T-t}\right) \boldsymbol{B}_2 & \cdots & \boldsymbol{B}_{T-1} \end{bmatrix}}_{\boldsymbol{S}_{\boldsymbol{u}}} \underbrace{\begin{bmatrix} \boldsymbol{u}_1 \\ \boldsymbol{u}_2 \\ \vdots \\ \boldsymbol{u}_{T-1} \end{bmatrix}}_{\boldsymbol{u}} \tag{2.2}$$

The above problem can be solved either with a batch formulation or a recursive formulation. The batch formulation computes open loop control commands, which can be used to solve a planning problem or to simulate the resulting trajectories in control space and in state space. The recursive formulation computes a controller with adaptive feedback gains that can, to a certain extent, handle external perturbations.

Batch formulation

With open loop control commands organized as a vector $\boldsymbol{u}$, the minimization of (2.1) subject to $\boldsymbol{x} = \boldsymbol{S}_{\boldsymbol{x}} \boldsymbol{x}_1 + \boldsymbol{S}_{\boldsymbol{u}} \boldsymbol{u}$ provides an analytic result given by

$$\hat{\boldsymbol{u}} = \left(\boldsymbol{S}_{\boldsymbol{u}}^\top \boldsymbol{Q} \boldsymbol{S}_{\boldsymbol{u}} + \boldsymbol{R}\right)^{-1} \boldsymbol{S}_{\boldsymbol{u}}^\top \boldsymbol{Q} \left(\boldsymbol{\mu} - \boldsymbol{S}_{\boldsymbol{x}} \boldsymbol{x}_1\right) \tag{2.3}$$

The residuals of this least squares solution provide information about the uncertainty of this estimate in the form of a full covariance matrix (at control trajectory level):

$$\hat{\Sigma}^u = \left(S_u^\top Q S_u + R\right)^{-1} \tag{2.4}$$

which provides a probabilistic interpretation of LQT. The resulting Gaussian distribution can then be projected back to a state trajectory space by exploiting the linear relationship between $\boldsymbol{u}$ and $\boldsymbol{x}$ in (2.2) so that a trajectory distribution in control space can be projected back to a trajectory distribution in state space. All the operations are analytic and only exploit basic linear algebra, see References 14 and 28 for details and Reference 29 for examples of codes.

Figure 2.5-*left* shows that an LQT cost is composed of quadratic terms, whose solution corresponds to a PoG in control trajectory space, where one Gaussian is centered on zero for all time steps (namely, staying still minimizes the control cost), and the other Gaussian corresponds to the swift tracking of a given reference. The PoG fuses the two constraints in control trajectory space. With this probabilistic interpretation, the batch formulation can be used to create bridges between learning and control. Figure 2.5-*center* shows an implementation example in LfD, where a set of demonstrations is encoded in a *hidden semi-Markov model* [30], providing an LQT cost function with the reference given by a stepwise reference built from the sequence of Gaussian centers and a full precision matrix built from the sequence of Gaussian covariances. This approach allows the observed (co)variations of a task to generate a controller with anticipation capability. Here, five demonstrations are encoded as three Gaussians. Without anticipation, the reproduced movement would move from one center to the other, without reproducing the Z shape of the demonstrations. Instead, LQT correctly reproduces the Z shape from any initial point $\boldsymbol{x}_0$ (see the trajectory produced by the controller in black) by exploiting the coordination patterns extracted from the demonstrations.

Figure 2.5-*right* highlights that the approach can be extended to *model predictive control*, *iterative LQR* (iLQR), and *differential dynamic programming*, whose solutions need this time either to be interpreted within an iterative process or updated at consecutive time steps, see Reference 31 for details.

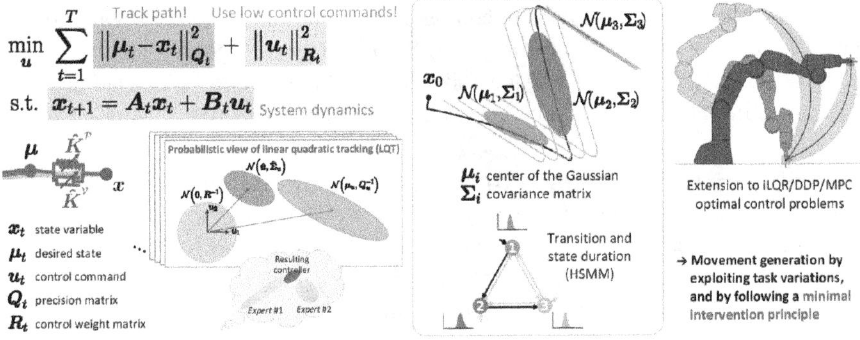

Figure 2.5 Probabilistic optimal control (see main text for details)

Recursive formulation

The problem of tracking a reference signal $\{\boldsymbol{\mu}_t\}_{t=1}^{T}$ can be recast as a regulation problem by considering a dynamical system $\tilde{\boldsymbol{x}}_{t+1} = \tilde{\boldsymbol{A}}_t\tilde{\boldsymbol{x}}_t + \tilde{\boldsymbol{B}}_t\boldsymbol{u}_t$ with an augmented state

$$\underbrace{\begin{bmatrix} \boldsymbol{x}_{t+1} \\ 1 \end{bmatrix}}_{\tilde{x}_{t+1}} = \underbrace{\begin{bmatrix} \boldsymbol{A}_t & \boldsymbol{0} \\ \boldsymbol{0} & 1 \end{bmatrix}}_{\tilde{A}_t} \underbrace{\begin{bmatrix} \boldsymbol{x}_t \\ 1 \end{bmatrix}}_{\tilde{x}_t} + \underbrace{\begin{bmatrix} \boldsymbol{B}_t \\ \boldsymbol{0} \end{bmatrix}}_{\tilde{B}_t} \boldsymbol{u}_t$$

and augmented tracking precision matrices

$$\tilde{\boldsymbol{Q}}_t = \begin{bmatrix} \boldsymbol{Q}_t^{-1} + \boldsymbol{\mu}_t\boldsymbol{\mu}_t^{\top} & \boldsymbol{\mu}_t \\ \boldsymbol{\mu}_t^{\top} & 1 \end{bmatrix}^{-1} = \begin{bmatrix} \boldsymbol{I} & \boldsymbol{0} \\ -\boldsymbol{\mu}_t^{\top} & 1 \end{bmatrix} \begin{bmatrix} \boldsymbol{Q}_t & \boldsymbol{0} \\ \boldsymbol{0} & 1 \end{bmatrix} \begin{bmatrix} \boldsymbol{I} & -\boldsymbol{\mu}_t \\ \boldsymbol{0} & 1 \end{bmatrix}$$

where the tilde notation $\tilde{\cdot}$ is used to refer to an augmented variable.

This augmented form is used to express the cost in (2.1) as

$$c = \tilde{\boldsymbol{x}}_T^{\top}\tilde{\boldsymbol{Q}}_T\tilde{\boldsymbol{x}}_T + \sum_{t=1}^{T-1} \tilde{\boldsymbol{x}}_t^{\top}\tilde{\boldsymbol{Q}}_t\tilde{\boldsymbol{x}}_t + \boldsymbol{u}_t^{\top}\boldsymbol{R}_t\boldsymbol{u}_t \tag{2.5}$$

which has the same form as a standard LQR problem. For a tracking problem, the resulting optimal control policy takes the form

$$\hat{\boldsymbol{u}}_t = -\tilde{\boldsymbol{K}}_t\,\tilde{\boldsymbol{x}}_t = \boldsymbol{K}_t(\boldsymbol{\mu}_t - \boldsymbol{x}_t) + \boldsymbol{u}_t^{\text{ff}} \tag{2.6}$$

characterized by a feedback gain matrix $\boldsymbol{K}_t$ extracted from $\tilde{\boldsymbol{K}}_t = [\boldsymbol{K}_t, \boldsymbol{k}_t]$, and a feedforward term $\boldsymbol{u}_t^{\text{ff}} = -\boldsymbol{k}_t - \boldsymbol{K}_t\boldsymbol{\mu}_t$ depending on $\boldsymbol{\mu}_t$.

The augmented feedback gain matrices in (2.6) are computed with

$$\tilde{\boldsymbol{K}}_t = (\tilde{\boldsymbol{B}}_t^{\top}\boldsymbol{P}_{t+1}\tilde{\boldsymbol{B}}_t + \boldsymbol{R}_t)^{-1}\tilde{\boldsymbol{B}}_t^{\top}\boldsymbol{P}_{t+1}\tilde{\boldsymbol{A}}_t \tag{2.7}$$

where $\boldsymbol{P}_t$ is evaluated recursively from $t = T-1$ to $t = 1$ as

$$\boldsymbol{P}_t = \tilde{\boldsymbol{A}}_t^{\top}\boldsymbol{P}_{t+1}\tilde{\boldsymbol{A}}_t + \tilde{\boldsymbol{Q}}_t\tilde{\boldsymbol{A}}_t^{\top}\boldsymbol{P}_{t+1}\tilde{\boldsymbol{B}}_t(\tilde{\boldsymbol{B}}_t^{\top}\boldsymbol{P}_{t+1}\tilde{\boldsymbol{B}}_t + \boldsymbol{R}_t)^{-1}\tilde{\boldsymbol{B}}_t^{\top}\boldsymbol{P}_{t+1}\tilde{\boldsymbol{A}}_t$$

starting from $\boldsymbol{P}_T = \tilde{\boldsymbol{Q}}_T$, see References 28 and 29 for details.

After all feedback gain matrices $\tilde{\boldsymbol{K}}_t$ have been computed by backward recursion, a forward recursion is used to compute the evolution of the state with the linear system $\tilde{\boldsymbol{x}}_{t+1} = \tilde{\boldsymbol{A}}_t\tilde{\boldsymbol{x}}_t + \tilde{\boldsymbol{B}}_t\hat{\boldsymbol{u}}_t$, starting from $\tilde{\boldsymbol{x}}_1$, by using the feedback control policy $\boldsymbol{u}_t = -\tilde{\boldsymbol{K}}_t\tilde{\boldsymbol{x}}_t$. Compared with a batch formulation, this recursive formulation computes a controller that can, to a certain extent, handle external perturbations.

Lease square computation of the recursive formulation

We will now show that an intermediary result exists between a batch and a recursive formulation and that this formulation is helpful to link learning and control algorithms. First, by using augmented vectors to describe trajectories and writing $\boldsymbol{u} = -\boldsymbol{F}\tilde{\boldsymbol{x}}_1$, we can redefine the costw in (2.1) and (2.5) as the optimization problem

$$\min_{\boldsymbol{F}} \; \tilde{\boldsymbol{x}}^{\top}\tilde{\boldsymbol{Q}}\tilde{\boldsymbol{x}} + (-\boldsymbol{F}\tilde{\boldsymbol{x}}_1)^{\top}\boldsymbol{R}(-\boldsymbol{F}\tilde{\boldsymbol{x}}_1), \quad \text{s.t.} \quad \tilde{\boldsymbol{x}} = (\tilde{\boldsymbol{S}}_x - \tilde{\boldsymbol{S}}_u\boldsymbol{F})\tilde{\boldsymbol{x}}_1$$

whose least squares solution is

$$F = (\tilde{S}_u^\top \tilde{Q} \tilde{S}_u + R)^{-1} \tilde{S}_u^\top \tilde{Q} \tilde{S}_x \tag{2.8}$$

which can be used to iteratively reconstruct regulation gains $\tilde{K}_t$ by starting from $\tilde{K}_1 = F_1, P_1 = I$. This is done by computing recursively:

$$P_t = P_{t-1}(\tilde{A}_{t-1} - \tilde{B}_{t-1}\tilde{K}_{t-1})^{-1}, \qquad \tilde{K}_t = F_t P_t \tag{2.9}$$

which can then be used in a feedback controller on an augmented state as in (2.6).

This least squares formulation of recursive LQT, that we will call lease square computation of the recursive formulation (LQT-LS), yields the same controller (2.6) as with the standard recursive formulation. However, the linear form in (2.8) used by LQT-LS has several advantages. First, it allows the use of full precision matrices $\tilde{Q}$ that can encode correlation constraints within a given state but also between time steps, see Reference 32 for an example. It also allows LQT-LS to be extended to the use of CP, which we discuss next.

LQT-LS with CP

A univariate control trajectory $u^{1D} \in \mathbb{R}^T$ of T control commands can be represented as a weighted sum of K basis functions with

$$u^{1D} = \sum_{k=1}^{K} \phi_k \, w_k^{1D} = \phi w^{1D} \tag{2.10}$$

where the matrix ϕ is a horizontal concatenation of univariate basis functions.

A multivariate control trajectory $u^{1D} \in \mathbb{R}^{DT}$ of T control commands of dimension D can similarly be computed as

$$u = \sum_{k=1}^{K} \Psi_k w_k = \Psi w, \text{ with } \Psi = \phi \otimes C = \begin{bmatrix} C\phi_{1,1} & C\phi_{2,1} & \cdots & C\phi_{K,1} \\ C\phi_{1,2} & C\phi_{2,2} & \cdots & C\phi_{K,2} \\ \vdots & \vdots & \ddots & \vdots \\ C\phi_{1,T} & C\phi_{2,T} & \cdots & C\phi_{K,T} \end{bmatrix} \tag{2.11}$$

where $\otimes$ is the Kronecker product operator, and C is a coordination matrix that can be set to an identity matrix I_D for the generic case of control variables with independent basis functions for each dimension $d \in \{1, ..., D\}$.

Equation (2.11) aims at encoding the control commands as a weighted superposition of simpler commands. In an optimal control problem, the encoding of control commands as a set of weights has two advantages: (1) it reduces computation time and storage by working in a subspace of reduced dimensionality; and (2) it acts as a regularization approach by denoising the signal while capturing the essential aspects of a controlled movement. Such a representation can be illustrated as the use of basic building blocks that can be assembled in different manners to form more elaborated movements, often referred to as MP.

If we assume that the control commands profile $\boldsymbol{u}$ is composed of CP with $\boldsymbol{u} = \boldsymbol{\Psi w}$, the LQT objective in (2.1) becomes

$$\min_{\boldsymbol{w}} (\boldsymbol{x} - \boldsymbol{\mu})^{\top} \boldsymbol{Q} (\boldsymbol{x} - \boldsymbol{\mu}) + \boldsymbol{w}^{\top} \boldsymbol{\Psi}^{\top} \boldsymbol{R} \boldsymbol{\Psi} \boldsymbol{w}, \quad \text{s.t.} \quad \boldsymbol{x} = \boldsymbol{S}_x \boldsymbol{x}_1 + \boldsymbol{S}_u \boldsymbol{\Psi} \boldsymbol{w}$$

with a solution given by

$$\hat{\boldsymbol{w}} = \left(\boldsymbol{\Psi}^{\top} \boldsymbol{S}_u^{\top} \boldsymbol{Q} \boldsymbol{S}_u \boldsymbol{\Psi} + \boldsymbol{\Psi}^{\top} \boldsymbol{R} \boldsymbol{\Psi}\right)^{-1} \boldsymbol{\Psi}^{\top} \boldsymbol{S}_u^{\top} \boldsymbol{Q} \left(\boldsymbol{\mu} - \boldsymbol{S}_x \boldsymbol{x}_1\right)$$

which is used to compute the trajectory $\hat{\boldsymbol{u}} = \boldsymbol{\Psi} \hat{\boldsymbol{w}}$ in control space, corresponding to the list of control commands organized in vector form. Note that in practice, $\boldsymbol{R}$ can be set to zero because the regularization role of $\boldsymbol{R}$ in the original problem formulation can be considered as redundant to the use of sparse basis functions.

Similarly to (2.4), the residuals of the least squares solution provide a probabilistic representation of LQT with CP. Note that as seen before, the trajectory in control space can be converted to a trajectory in state space with $\hat{\boldsymbol{x}} = \boldsymbol{S}_x \boldsymbol{x}_1 + \boldsymbol{S}_u \hat{\boldsymbol{u}}$. Thus, since all operations are linear, a covariance matrix can be converted to a covariance on $\boldsymbol{u}$ and on $\boldsymbol{x}$. A similar linear transformation is used in the context of *probabilistic* MP [33] to map a covariance matrix on $\boldsymbol{w}$ to a covariance on $\boldsymbol{x}$. LQT with CP provides a more general approach by considering both control and state spaces. It also has the advantage of reframing the method to the more general context of optimal control, thus providing various extension opportunities.

While radial basis functions (RBFs) are the most commonly used representation in MP, other forms of basis functions can be considered, including stepwise control commands, Bernstein polynomials, B-splines, or Fourier series, which can all be organized as a dictionary matrix $\boldsymbol{\Psi}$. These basis functions can also be optimized instead of predefined [34].

Figure 2.6 presents some examples, see Reference 35 for details. The use of an exponential in RBFs makes them infinitely differentiable. However, they typically require a bandwidth parameter to be set heuristically. Bernstein basis functions and Fourier basis functions do not require this type of hyperparameter to be set. Bernstein basis functions instead require the degree of the polynomial to be set, which can be selected based on the number of derivatives required. Their advantages are that they can be interpreted as control points (corresponding to Bézier curves) and that they sum to a regular constant signal. Fourier basis functions are infinitely differentiable and provide a spectral decomposition of the signal instead of a spatial decomposition, with an ordering of the basis functions corresponding to the resolution required

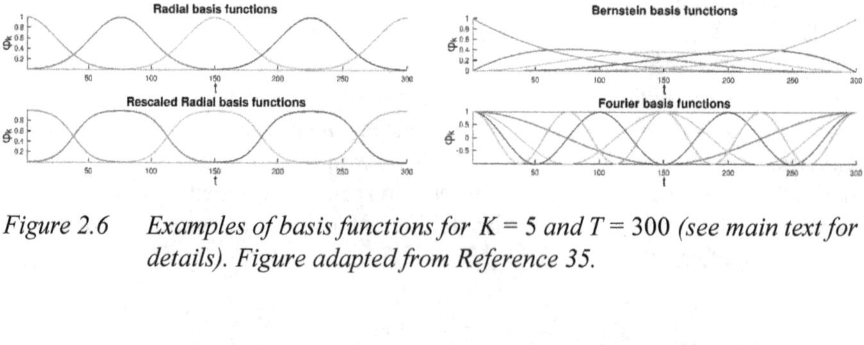

Figure 2.6 Examples of basis functions for $K = 5$ *and* $T = 300$ *(see main text for details). Figure adapted from Reference 35.*

for the reconstruction of the signal. They can be used for both periodic and discrete motion by encoding a mirrored version of the original signal.

Next, we present an example of the proposed LQT-LS formulation with basis functions to implement the DMP principle [36].

2.4.2.2 Dynamical movement primitives (DMP) formulated as LQT-LS with control primitives

The DMP [36] approach proposes the reproduction of an observed movement by considering a controller composed of two parts: a closed-loop spring-damper system reaching the final point of the observed movement and an open-loop forcing term composed of the acceleration, velocity, and position profiles of the observed movement. The evolution of this system is either driven by time or by a decay term. These two controllers are weighted so that the influence of the spring-damper system is progressively increased until it becomes the only active controller. In DMP, the forcing term is encoded with RBFs [37]. The spring-damper system parameters of the closed-loop system are defined heuristically, usually as a critically damped system, see Reference 36 for details.

The LQT-LS approach presented in the previous section can be employed to implement the DMP principle, with a linear system defined as an integrator (e.g., second-order integrator). The LQT cost can similarly request a target to be reached at the end of the movement, while following the observed acceleration/velocity/position profiles. The controller can be estimated either as open-loop control commands (2.3) (providing a probabilistic interpretation) or as a closed-loop controller (2.6) by encoding $\boldsymbol{F}$ in (2.8) as RBFs.

In the latter case, the matrix $\boldsymbol{F}$ in (2.8) is estimated by using CP and an augmented state space formulation, namely

$$\hat{\boldsymbol{W}} = \left(\boldsymbol{\Psi}^{\top}\tilde{\boldsymbol{S}}_u^{\top}\tilde{\boldsymbol{Q}}\tilde{\boldsymbol{S}}_u\boldsymbol{\Psi} + \boldsymbol{\Psi}^{\top}\boldsymbol{R}\boldsymbol{\Psi}\right)^{-1}\boldsymbol{\Psi}^{\top}\tilde{\boldsymbol{S}}_u^{\top}\tilde{\boldsymbol{Q}}\tilde{\boldsymbol{S}}_x, \qquad \boldsymbol{F} = \boldsymbol{\Psi}\hat{\boldsymbol{W}}$$

which is used to compute feedback gains $\tilde{\boldsymbol{K}}_t$ on the augmented state with (2.9).

The resulting feedback controller $\hat{\boldsymbol{u}} = -\tilde{\boldsymbol{K}}_t\tilde{\boldsymbol{x}}_t$ tracks the observed profiles while smoothly reaching the desired goal at the end of the movement, with a smooth transition between the two control objectives. The main difference with DMP is that the smooth transition between the two behaviors is directly optimized by the system, together with the parameters of the feedback controller that are also automatically optimized (including stiffness and damping ratio).

The proposed LQT-LS formulation of DMP allows the resulting controller to be formalized as a cost function, which allows the approach to be combined more easily with other optimal control strategies. First, the proposed approach can freely modulate the way in which the system reacts to perturbations by weighting how important it is to follow the demonstrated position, velocity, or acceleration profiles, respectively (or higher order derivatives, if desired). For example, setting a higher cost on the position profile will make the feedback controller more rapidly move back to the path instead of following the velocity and acceleration profiles. Moreover, the approach can be extended to multiple viapoints without any modification. It also

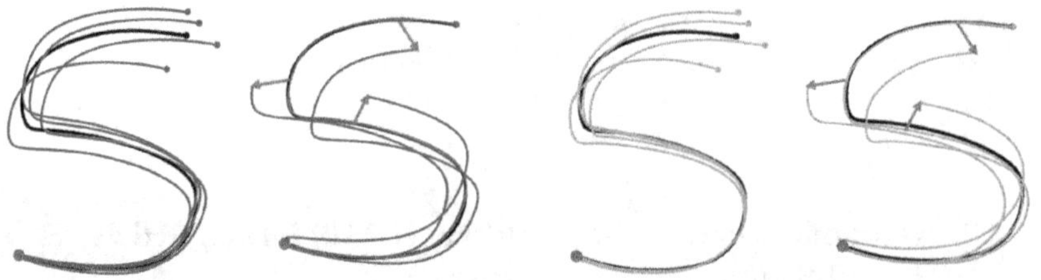

(a) Proposed LQT-LS (b) Standard DMP

Figure 2.7 *Least squares solution of recursive linear quadratic tracking (LQT-LS) by using CP. To imitate a DMP, the cost of LQT-LS can be set to track both the demonstrated motion (in black) and the endpoint (in red). In this example, LQT-LS and DMP both use 15 RBFs. The two models (LQT-LS in blue, DMP in green) are tested with the same set of four random initial states and four random perturbations (orange arrow) occurring during the motion, showing the similarities of the two results.*

allows multiple demonstrations of a movement to be used to estimate the feedback controller, which provides a probabilistic formulation of DMP. As a result, the (co) variations in the demonstrations are exploited to provide a minimal intervention control strategy that will selectively reject perturbations, based on the impact that these perturbations have on the task to achieve. This is effectively attained by automatically regulating the gains in accordance to the variations in the demonstrations, with low gains in parts of the movement allowing variations and higher gains for parts of the movement that are invariant in the demonstrations.

Figure 2.7 presents an example by reproducing a demonstrated S-shaped trajectory, see Reference 29 for the corresponding code example.

2.5 Geometric structures

In the previous sections, we have discussed learning and control problems in standard Euclidean spaces. In this section, we extend the proposed challenges by adopting a geometric perspective.

Data encountered in robotics are characterized by simple but varied geometries. These geometric structures are often underexploited in learning, planning, control, and perception. Riemannian geometry provides a principled and simple way to extend algorithms initially developed for Euclidean data to other manifolds by efficiently taking into account prior geometric knowledge about these manifolds [38, 39]. In particular, Riemannian manifolds can be connected to a wide range of problems in robotics, including problems relying on Gaussian distributions [21]. This includes techniques requiring uncertainty and statistical modeling to be computed on structured non-Euclidean data.

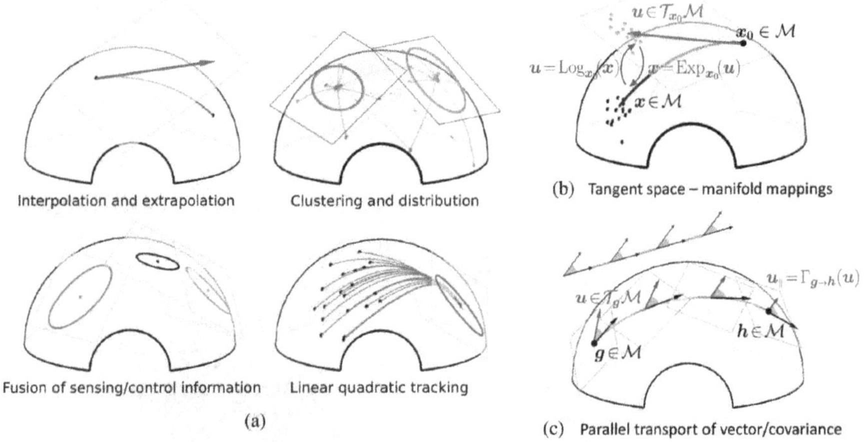

Figure 2.8 *(a) Riemannian geometry can solve a variety of problems involving distributions and uncertainties. Applications in robotics using Riemannian manifolds principally rely on two principles: (b) exponential/logarithmic mapping between tangent space and manifold; and (c) parallel transport of a vector along a geodesic (see main text for details).*

Examples of standard problems in robotics are depicted in Figure 2.8(a) from a geometric perspective. Riemannian manifolds can be applied to various forms of data, as illustrated in Figure 2.9. Such data range from joint angles in revolving articulations [40], rigid body motions [41, 42], unit quaternions to represent orientations [43], and symmetric positive definite (SPD) matrices, which can represent sensory data processed as spatial covariances [44], inertia [45, 46], stiffness/manipulability ellipsoids [47], as well as metrics used in the context of similarity measures.

By using Riemannian manifolds, data of various forms can be treated in a unified manner, with the advantage that existing models and algorithms initially developed for Euclidean data can be extended to a wider range of data structures. It can for example be used to revisit constrained optimization problems formulated in Euclidean space, with a treatment as unconstrained problems inherently taking into account the geometry of the data.

A d-dimensional *Riemannian manifold* $\mathcal{M}$ is a smooth and differentiable manifold, which locally behaves like the Euclidean space $\mathbb{R}^d$ and which is equipped with a positive definite metric tensor to measure distances of points on the manifold. For each point $\boldsymbol{p} \in \mathcal{M}$, there exists a tangent space $\mathcal{T}_p\mathcal{M}$ that locally linearizes the manifold. The affine connection, computed from the metric, is a differential operator that provides a way to compute geodesics and to transport vectors on tangent spaces along any smooth curves on the manifold [38, 39]. The Cartesian product of two Riemannian manifolds is also a Riemannian manifold, which allows joint distributions to be constructed with any combination of Riemannian manifolds.

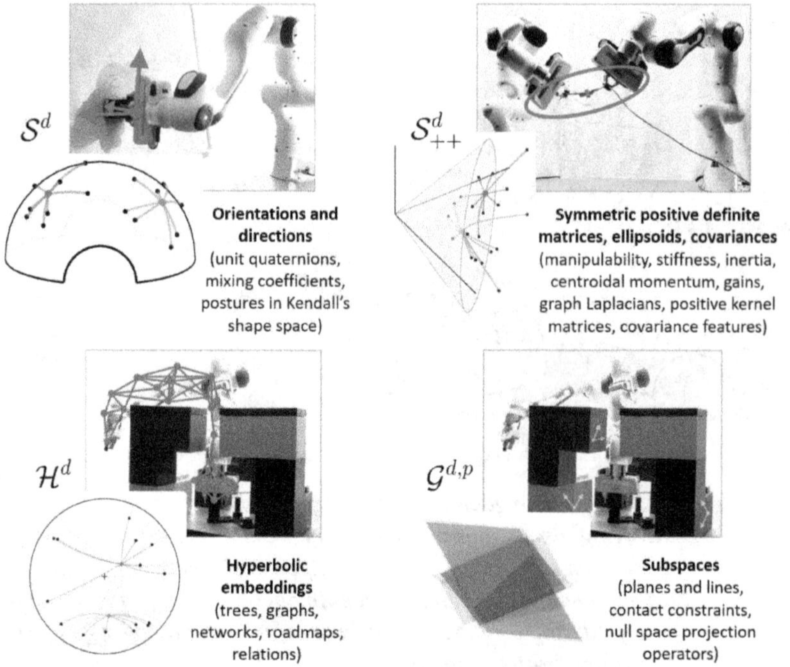

Figure 2.9 *Several geometric structures are relevant for industrial robotics applications. $\mathcal{S}^3$ can be used to represent the orientation of robot end effectors (unit quaternions). $\mathcal{S}^6_{++}$ can be used to represent manipulability ellipsoids (manipulability capability in translation and rotation), corresponding to SPD matrix manifolds. $\mathcal{H}^d$ can be used to represent trees, graphs, and roadmaps. $\mathcal{G}^{d,p}$ can be used to represent subspaces (planes, nullspaces, and projection operators).*

Two basic notions of Riemannian geometry are crucial for learning and control applications, which are illustrated in Figure 2.8(b, c) and described below.

Geodesics: The minimum length curves between two points on a Riemannian manifold are called geodesics. The exponential map $\mathrm{EXP}_{x_0}: \mathcal{T}_{x_0}\mathcal{M} \rightarrow \mathcal{M}$ maps a point $\boldsymbol{u}$ in the tangent space of $\boldsymbol{x}_0$ to a point $\boldsymbol{x}$ on the manifold so that $\boldsymbol{x}$ lies on the geodesic starting at $\boldsymbol{x}_0$ in the direction $\boldsymbol{u}$. The norm of $\boldsymbol{u}$ is equal to the geodesic distance between $\boldsymbol{x}_0$ and $\boldsymbol{x}$. The inverse map is called the logarithmic map $\mathrm{Log}_{x_0}: \mathcal{M} \rightarrow \mathcal{T}_{x_0}\mathcal{M}$. Figure 2.8(b) depicts these mapping functions.

Parallel transport: Parallel transport $\Gamma_{g\rightarrow h}: \mathcal{T}_g\mathcal{M} \rightarrow \mathcal{T}_h\mathcal{M}$ moves vectors between tangent spaces such that the inner product between two vectors in a tangent space is conserved. It employs the notion of connection, defining how to associate vectors between infinitesimally close tangent spaces. This connection allows the smooth transport of a vector from one tangent space to another.

Figure 2.8(c) depicts the parallel transport operation. The flat surfaces show the coordinate systems of several tangent spaces along the geodesic. The black vectors represent the directions of the geodesic in the tangent spaces. The blue vectors are transported from $\boldsymbol{g}$ to $\boldsymbol{h}$. Parallel transport allows a vector $\boldsymbol{u}$ in the tangent space of $\boldsymbol{g}$ to be transported to the tangent space of $\boldsymbol{h}$ by ensuring that the angle (i.e., inner product) between $\boldsymbol{u}$ and the direction of the geodesic (represented as black vectors) is conserved. At point $\boldsymbol{g}$, this direction is expressed as $\text{Log}_{\boldsymbol{g}}(\boldsymbol{h})$. This operation is crucial to combine information available at $\boldsymbol{g}$ with information available at $\boldsymbol{h}$ by taking into account the transformation of the coordinate systems along the geodesic (notice the rotation of the tangent spaces in Figure 2.8(c)). In Euclidean space (top-left inset in Figure 2.8(c)), parallel transport is simply the identity operator (a vector operation can be applied to any point without additional transformation).

By extension, a covariance matrix $\boldsymbol{\Sigma}$ can be transported by using a decomposition into eigenvectors, by transporting the eigenvectors, and by recomputing a covariance matrix with the transported eigenvectors. For many manifolds in robotics, this transport operation can be expressed as linear mapping. The most common manifolds in robotics are homogeneous, providing simple analytic expressions for exponential/logarithmic mapping and parallel transport. For industrial robotics, these manifolds include orientations, ellipsoids, graphs, and subspaces, see Figure 2.9.

Several approaches have been proposed to extend Gaussian distributions in Euclidean space to Riemannian manifolds. For robotics applications, a simple approach consists of estimating the mean of the Gaussian as a centroid on the manifold and representing the dispersion of the data as a covariance defined in the tangent space of the centroid. Parameters estimation can be solved by a simple and fast Gauss–Newton iterative algorithm. Importantly, this geometric mean can be directly extended to weighted distances, which can be exploited for mixture modeling, fusion (PoG), regression (conditional distribution), MP, and planning problems, see Figure 2.8(a) and Reference 21 for a review.

The combination of statistics and Riemannian geometry offers many research opportunities and can contribute to recent challenges in industrial robotics. Further work can be organized in two categories. First, the field of industrial robotics is abundant of new research developments due to the interdisciplinary aspect and to the richness of problems it involves (in particular, targeting cobots and bimanual manufacturing robots). The common factor in many of these developments is that they rely on some form of statistics and/or propagation of uncertainty. These models and algorithms are typically developed for standard Euclidean spaces, where an extension to Riemannian manifolds has several benefits to offer. Second, some Riemannian manifolds remain largely underexploited in robotics, despite the fact that some of them are mathematically well understood and characterized by simple closed-form expressions. Grassmann manifolds seem particularly promising to handle problems in robotics with high-dimensional datapoints and only few training data, where subspaces are required in the computation to keep the most essential characteristics of the data. It is also promising for problems in which hierarchies are considered (such as inverse kinematics with kinematically redundant robots) because it provides a geometric interpretation of nullspace structures. Other Riemannian manifolds such as hyperbolic manifolds also seem propitious to bring a probabilistic treatment to tree-based structures, graphs, Toeplitz/Hankel matrices, autoregressive models, and dynamical systems.

Next, we present an example exploiting Riemannian geometry for a robot application targeting skill transfer problems.

2.5.1 Learning, tracking, and transfer of manipulability ellipsoids

Stiffness and manipulability ellipsoids are geometric descriptors that can be used to transfer manipulation skills to a robot. As these ellipsoids lie on SPD manifolds, Riemannian manifolds can be used to learn and reproduce these descriptors in a probabilistic manner [21, 47].

Manipulability ellipsoids come in many different flavors due to the different types of Jacobian matrices that can be considered in robotics problems. They can be defined for different points of interest (end effector and center of mass), for positions and forces (including the orientation/rotational part), or for different task priorities (nullspace structures). Manipulability can be described at either kinematic or dynamic levels (including inertia), for either open or closed linkages (e.g., for bimanual manipulation or in-hand manipulation), and for either fully actuated or underactuated systems [48]. This large set of manipulability ellipsoids provides rich descriptors to characterize robot skills for robots with legs and arms.

Manipulability ellipsoids are helpful descriptors to handle skills transfer problems involving dissimilar kinematic chains, such as transferring manipulation skills between humans and robots or between two robots with different kinematic chains or capabilities. In such transfer problems, imitation at the joint angles level is not possible due to the different structures, and imitation at the end effector(s) level is limited because it does not encapsulate postural information, which is often an essential aspect of the skill that needs to be transferred. Manipulability ellipsoids provide intermediate descriptors that allow postural information to be transferred indirectly, with the advantage that they allow different embodiments and capabilities to be considered in the skill transfer problem.

Figure 2.10 presents an example within the context of industrial manipulation tasks to be transferred from a human user to a bimanual quadruped robot (centaur robot), see Reference 50 for the full description of the approach and experiments.

The manipulability ellipsoid $\boldsymbol{M}(\boldsymbol{x}) = \boldsymbol{J}(\boldsymbol{x})\,\boldsymbol{J}(\boldsymbol{x})^\top$ is an SPD matrix representing the manipulability at a given point on the kinematic chain defined by a forward kinematics function $\boldsymbol{f}(\boldsymbol{x})$, given the joint angle posture $\boldsymbol{x}$, where $\boldsymbol{J}(\boldsymbol{x})$ is the Jacobian of $\boldsymbol{f}(\boldsymbol{x})$. In robotics, the determinant of $\boldsymbol{M}(\boldsymbol{x})$ is most often employed [51]. It reduces the information to a scalar manipulability index indicating the volume of the ellipsoid, with the drawback of ignoring the specific shape of this ellipsoid. Such a scalar descriptor can have limitations when used as a metric to avoid kinematic singularity, see References 52 and 53.

In Reference 47, we showed that a geometric cost on manipulability can alternatively be defined with the geodesic distance

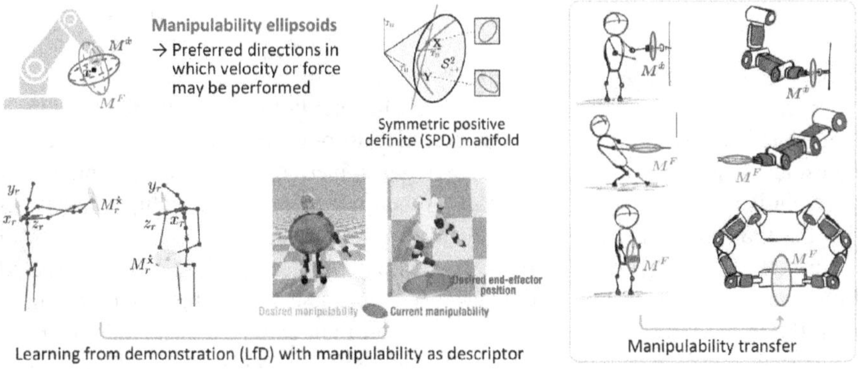

Figure 2.10 *Manipulability ellipsoids used as task descriptors, exploited in the context of transferring skills from humans to robots of different morphologies, by using the industrial operation dataset from Reference 49. Figure adapted from Reference 50.*

$$
\begin{aligned}
& c = \|\boldsymbol{A}\|_{\mathrm{F}}^{2}, \quad \text{with} \quad \boldsymbol{A} = \log\left(\boldsymbol{S}^{\frac{1}{2}} \boldsymbol{M}(\boldsymbol{x}) \boldsymbol{S}^{\frac{1}{2}}\right), \\
\iff \quad & c = \operatorname{trace}(\boldsymbol{A}\boldsymbol{A}^{\top}) = \sum_i \operatorname{trace}(\boldsymbol{A}_i\boldsymbol{A}_i^{\top}) = \sum_i \boldsymbol{A}_i^{\top}\boldsymbol{A}_i = \operatorname{vec}(\boldsymbol{A})^{\top}\operatorname{vec}(\boldsymbol{A})
\end{aligned}
\tag{2.12}
$$

where $\boldsymbol{S}$ is the desired manipulability matrix to reach, $\log(\cdot)$ is the logarithm matrix function, and $\|\cdot\|_{\mathrm{F}}$ is a Frobenius norm. By exploiting the trace properties, (2.12) can be expressed in a quadratic form, where a vectorization operation $\operatorname{vec}(\boldsymbol{A})$ can be computed efficiently by exploiting the symmetry of the matrix $\boldsymbol{A}$. This is implemented by keeping only the lower half of the matrix, with the elements below the diagonal multiplied by a factor $\sqrt{2}$.

The derivatives of (2.12) with respect to the state $\boldsymbol{x}$ and control commands $\boldsymbol{u}$ can be computed numerically by automatic differentiation or by following an analytic approach as in Reference 47. The quadratic form proposed in (2.12) can be exploited to solve an optimal control problem in the form of iLQR, formulated as a Gauss–Newton optimization problem leveraging the Jacobian of $\operatorname{vec}(\boldsymbol{A})$, see Reference 29 for a corresponding code example. Marić *et al.* demonstrated the advantages of a geometric cost as in (2.12) for planning problems by comparing it with other widely used metrics such as the manipulability index and the dexterity index [53].

As mentioned above, manipulability can be defined at several points of interest, including end effectors and centers of mass. It can also be defined at specific points on objects or tools held by the robot, allowing manipulability ellipsoids to be exploited as useful descriptors for object affordances. For example, for a robot holding a hammer, we can consider a manipulability ellipsoid at the head of a hammer. When the robot grasps the hammer, its end effector is extended so that the head of the hammer becomes the extremity of the kinematic chain. In this situation, the different options that the robot has to grasp the hammer will have an impact on the resulting manipulability at the head of the hammer. Thus, grabbing the hammer at the extremity of the handle will improve the resulting manipulability. Such descriptors offer promises for learning and optimization in manufacturing environments in order to let the robots automatically determine the correct ways to employ tools, including grasping points and body postures.

Optimal control with a geodesic cost on SPD manifolds can be used for manipulability ellipsoids, but it can also be extended to other descriptors represented as ellipsoids. In particular, stiffness, feedback gains, inertia, and centroidal momentum have similar structures. For the latter, the centroidal momentum ellipsoid quantifies the momentum generation ability of the robot [54], which is another useful descriptor for robot skills. Similarly to manipulability ellipsoids, it is constructed from Jacobians, which are in this case the centroidal momentum matrices mapping the joint angle velocities to the centroidal momentum, which sums over the individual link momenta after projecting each to the robot's center of mass. The proposed approach could also be extended to other forms of SPD matrices, such as kernel matrices used in machine learning algorithms to compute similarities, or graph Laplacians (a matrix representation of a graph that can, for example, be used to construct a low dimensional embedding of a graph).

2.6 Data structures

Another type of structures that can be exploited in learning and control problems in robotics relates to the organization of data as multidimensional arrays (also called tensors). These data appear in various robotics tasks, either as the natural organization of sensory/motor data (tactile arrays, images, and kinematic chains), as the result of standardized preprocessing steps (moving time windows and covariance features), as data in multiple coordinate systems, or in the form of basis functions decomposition.

Developed in the fields of *multilinear algebra* and *tensor methods*, tensor factorization techniques extend matrix factorization techniques such as singular value decomposition to multilinear decomposition, without requiring the transformation of the tensors to vectors or matrices. Figure 2.11 presents a few examples of tensor factorization techniques. These techniques can be exploited to provide robots with the capability to learn tasks from only few human demonstrations [55] or from only few exploration trials [56] by effectively leveraging the multidimensional nature of the problem.

Next, we present an example of tensor factorization used in the context of industrial peg-in-hole insertion tasks.

2.6.1 Ergodic control for insertion tasks

A tracking problem in industrial robot application is conventionally characterized by a target point to reach, requiring a controller to be computed to reach this target. In ergodic control, instead of providing a single target point, a probability distribution is given to the robot, which must cover this distribution in an efficient way. *Ergodic control* thus consists of moving within a spatial distribution by spending time in each part of the distribution in proportion to its density (illustratively, "tracking a

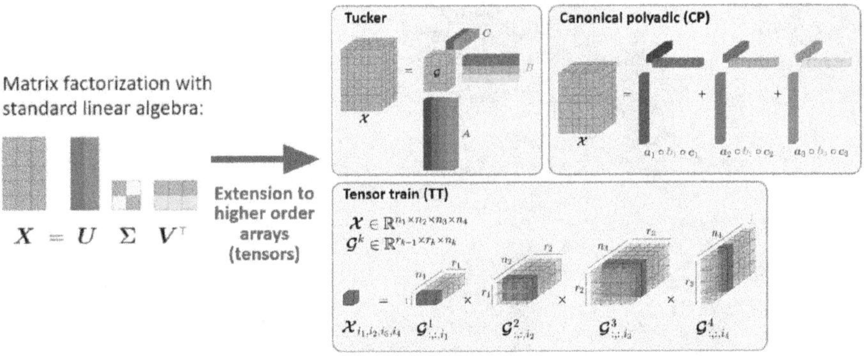

Figure 2.11 *Tensor factorization structures that can be used to automatically extract essential information from multidimensional arrays of data*

distribution" instead of "tracking a point"). The resulting controller generates natural exploration behaviors. The underlying objective takes a simple form, corresponding to a tracking problem in the spectral domain (tracking of frequency components). The advantage of such a control formulation is that it can be easily combined with other control objectives and constraints.

For industrial robot applications, the approach can be exploited in a wide range of problems and results in the automatic exploration of regions of interest. This is particularly helpful when the available sensing information is not accurate enough to fulfill the task with a standard controller, but where this information can still guide the robot toward promising areas. In a collaborative task, it can also be used when the operator's input is not accurate enough to fully reproduce the task, which then requires the robot to explore around the requested input (e.g., a point of interest selected by the operator). For picking and insertion problems, ergodic control can be applied to move around the picking/insertion point, thereby facilitating the prehension/entry. It can also be employed for active sensing and localization (either detected autonomously or with help by the operator). Here, the robot can plan movements based on the current information density. It can then recompute the commands when new measurements are available (i.e., updating the spatial distribution used as a target).

In ergodic control, the problem is originally formulated as a *spectral multiscale coverage* objective [57]. It requires the spatial distribution to be decomposed as Fourier series, with a cost function comparing the spectral decomposition of the robot path with the spectral decomposition of the distribution. The resulting controller allows the robot to explore the given spatial distribution in a natural manner by starting from a crude exploration and by refining the search progressively (i.e., matching the Fourier coefficients with increasing importance from low to high-frequency components).

Figure 2.12 presents a 2D example. Figure 2.12(a) shows the spatial distribution $\hat{g}(x)$ that the agent has to explore, encoded here as a mixture of two Gaussians (gray colormap in left graph). The right graphs show the corresponding Fourier

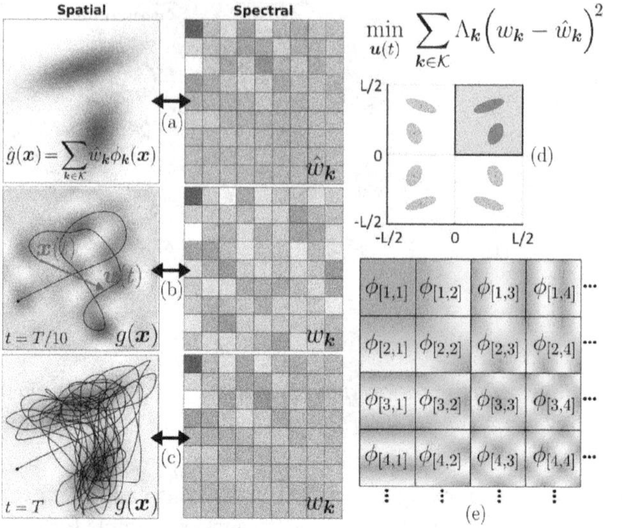

Figure 2.12 *2D ergodic control used to generate search behaviors (see main text for details). Figure adapted from Reference 35.*

series coefficients $\hat{w}_k$ in the frequency domain ($K = 9$ coefficients per dimension). Figure 2.12(b) shows the evolution of the reconstructed spatial distribution $g(\boldsymbol{x})$ (left graph) and the computation of the next control command $\boldsymbol{u}$ (red arrow) after $T/10$ iterations. The corresponding Fourier series coefficients w_k are shown in the right graph. Figure 2.12(c) shows that after T iterations, the agent covers the space in proportion to the desired spatial distribution, with a good match of coefficients in the frequency domain (we can see that $\hat{w}_k$ and w_k are nearly the same). Figure 2.12(d) shows how a periodic signal $\hat{g}(\boldsymbol{x})$ (with range $[-L/2, L/2]$ for each dimension) can be constructed from an initial mixture of two Gaussians $\hat{g}_0(\boldsymbol{x})$ (red area). The constructed signal $\hat{g}(\boldsymbol{x})$ is composed of eight Gaussians in this 2D example by mirroring the Gaussians along horizontal and vertical axes to construct an even signal of period L. Figure 2.12(e) depicts the first few basis functions of the Fourier series for the first four coefficients in each dimension, represented as a 2D colormap corresponding to periodic signals of different frequencies along two axes.

The downside of the original ergodic control approach is that it does not scale well to problems requiring exploration in search space of more than 2D. In particular, the original approach would be too slow to consider full distributions in 6D spaces, which would ideally be required. Indeed, both position and orientation of end effector(s) matter in most industrial robot problems, including manipulation, insertion, welding, or the use of tools at the robot end effectors. For this reason, alternative approaches mimicking similar behaviors, but departing from the original problem formulation, have been proposed [58, 59].

In Reference 60, we demonstrated that the original problem formulation of Reference 57 can be conserved by compressing the Fourier series decomposition

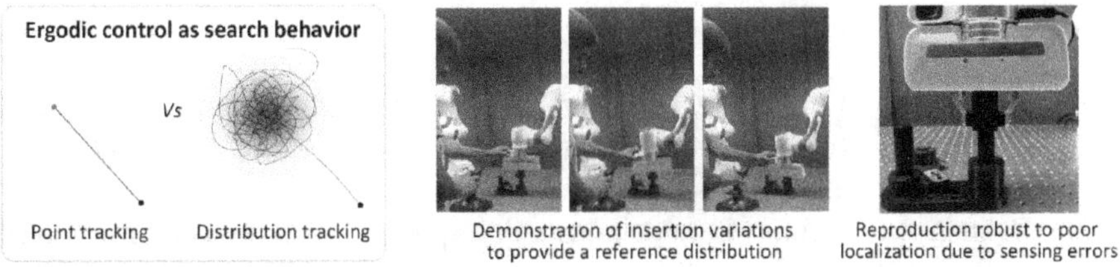

Figure 2.13 *Insertion tasks with ergodic control by using the Siemens gears benchmark. Figure adapted from Reference 60.*

with *tensor train* (TT) factorization. The proposed solution is efficient both computationally and storage-wise, hence making it suitable for online implementations, as well as to tackle robot learning problems with a low quantity of training data. The proposed approach using TTs and Fourier series to compress the spatial distributions has been tested in various experiments.

Figure 2.13 shows an overview of the insertion experiment with the Siemens gears benchmark, requiring full 6D end effector poses, see Reference 60 for details of the method and results. The experiment showed that the proposed approach can easily handle 6D spatial coverage problems (including position and orientation), with fast computation and small memory requirements.

2.7 Conclusion

This chapter presented an overview of the research directions that are currently investigated in the Robot Learning and Interaction group at the Idiap Research Institute to facilitate robot LfD. In contrast to the majority of ongoing work in machine learning, our research targets small data learning problems, with the goal of reducing the number of demonstrations that an operator has to provide to a robot in order to let it acquire manipulation skills that can generalize to new situations. This objective led to the consideration of a diverse set of structures, including learning strategy structures, combination structures, geometric structures, and tensor data structures. This chapter showed how these structures can be beneficial to robot skill acquisition problems, with the goal of reducing the amount of data required to learn manipulations tasks, and of generalizing the learned tasks to new situations, to new robots, and to new users. An overview of each approach was presented in the context of industrial tasks by discussing research directions that still need to be investigated to enable further technology transfer to industrial robots.

Acknowledgments

This work was supported by the CoLLaboratE project (https://collaborate-project.eu), funded by the EU within H2020-DTFOF-02-2018 under grant agreement 820767.

References

[1] Billard A.G., Calinon S., Dillmann R. 'Learning from humans' in Siciliano B., Khatib O. (eds.). *Handbook of Robotics*. 2nd edition. Secaucus, NJ, USA: Springer; 2016. pp. 1995–2014.

[2] Calinon S. 'Learning from demonstration (programming by demonstration)' in Ang M.H., Khatib O., Siciliano B. (eds.). *Encyclopedia of Robotics*. Berlin, Heidelberg: Springer; 2019.

[3] Chatzilygeroudis K., Vassiliades V., Stulp F., Calinon S., Mouret J.-B. 'A survey on policy search algorithms for learning robot controllers in a handful of trials'. *IEEE Transactions on Robotics*. 2020, vol. 32(2), pp. 328–47.

[4] Thomaz A.L., Breazeal C. 'Teachable robots: understanding human teaching behavior to build more effective robot learners'. *Artificial Intelligence*. 2008, vol. 172(6–7), pp. 716–37.

[5] Cakmak M., DePalma N., Arriaga R.I., Thomaz A.L. 'Exploiting social partners in robot learning'. *Autonomous Robots*. 2010, vol. 29(3–4), pp. 309–29.

[6] Kulak T., Calinon S. 'Combining social and intrinsically-motivated learning for multi-task robot skill acquisition'. IEEE Transactions on Cognitive and Developmental Systems. 2021.

[7] Papageorgiou D., Dimeas F., Kastritsi T., Doulgeri Z. 'Kinesthetic guidance utilizing DMP synchronization and assistive virtual fixtures for progressive automation'. *Robotica*. 2020, vol. 38(10), pp. 1824–41.

[8] Makhataeva Z., Varol H.A. 'Augmented reality for robotics: a review'. *Robotics*. 2020, vol. 9(2), p. 21.

[9] Thoo Y.J., Maceiras J., Abbet P, *et al.* 'Online and offline robot programming via augmented reality workspaces'. *arXiv:210701884*. 2021, pp. 1–8.

[10] Kimble K., Van Wyk K., Falco J., *et al.* 'Benchmarking protocols for evaluating small parts robotic assembly systems'. *IEEE Robotics and Automation Letters*. 2020, vol. 5(2), pp. 883–89.

[11] Mussa-Ivaldi F.A., Giszter S.F., Bizzi E. 'Linear combinations of primitives in vertebrate motor control'. *Proceedings of the National Academy of Sciences*. 1994, vol. 91(16), pp. 7534–38.

[12] Pignat E., Silvério J., Calinon S. 'Learning from demonstration using products of experts: applications to manipulation and task prioritization'. *The International Journal of Robotics Research*. 2022, vol. 41(2), pp. 163–88.

[13] Hinton G.E. 'Products of experts'. *Proc. Intl Conf. on Artificial Neural Networks (ICANN)*; 1999. pp. 1–6.

[14] Calinon S. 'Stochastic learning and control in multiple coordinate systems' in *Intl workshop on human-friendly robotics*. Genova, Italy; 2016. pp. 1–5.

[15] Kelso JAS. 'Synergies: atoms of brain and behavior' in Sternad D. (ed.). Progress in motor control. Vol. 629 of advances in experimental medicine and biology. Springer US; 2009. pp. 83–91.

[16] Todorov E., Jordan MI. 'Optimal feedback control as a theory of motor coordination'. *Nature Neuroscience*. 2002, vol. 5(11), pp. 1226–35.

[17] Scholz J.P., Schöner G. 'The uncontrolled manifold concept: identifying control variables for a functional task'. *Experimental Brain Research*. 1999, vol. 126(3), pp. 289–306.

[18] Wolpert D.M., Diedrichsen J., Flanagan J.R. 'Principles of sensorimotor learning'. *Nature Reviews. Neuroscience*. 2011, vol. 12, pp. 739–51.

[19] Calinon S. 'A tutorial on task-parameterized movement learning and retrieval'. *Intelligent Service Robotics*. 2016, vol. 9(1), pp. 1–29.

[20] Giusti A., Zeestraten M.J.A., Icer E., *et al.* 'Flexible automation driven by demonstration: leveraging strategies that simplify robotics'. *IEEE Robotics & Automation Magazine (RAM)*. 2018, vol. 25(2), pp. 18–27.

[21] Calinon S. 'Gaussians on riemannian manifolds: applications for robot learning and adaptive control'. *IEEE Robotics & Automation Magazine*. 2020, vol. 27(2), pp. 33–45.

[22] Birk A., Antonelli G., Di Lillo P, *et al.* 'Dexterous underwater manipulation from onshore locations: streamlining efficiencies for remotely operated underwater vehicles'. *IEEE Robotics & Automation Magazine (RAM)*. 2018, vol. 25(4), pp. 24–33.

[23] Ti B., Gao Y., Zhao J., Calinon S. 'Imitation of manipulation skills using multiple geometries'. *2022 IEEE/RSJ International Conference on Intelligent Robots and Systems (IROS)*; Kyoto, Japan, 2022.

[24] Silvério J., Calinon S., Rozo L., Caldwell D.G. 'Learning task priorities from demonstrations'. *IEEE Transactions on Robotics*. 2019, vol. 35(1), pp. 78–94.

[25] Whiten A., McGuigan N., Marshall-Pescini S., Hopper L.M. 'Emulation, imitation, over-imitation and the scope of culture for child and chimpanzee'. *Philosophical Transactions of the Royal Society of London. Series B, Biological Sciences*. 2009, vol. 364(1528), pp. 2417–28.

[26] Akgun B., Thomaz A. 'Simultaneously learning actions and goals from demonstration'. *Autonomous Robots*. 2016, vol. 40(2), pp. 211–27.

[27] Razmjoo A., Lembono T.S., Calinon S. 'Optimal control combining emulation and imitation to acquire physical assistance skills'. *2021 20th International Conference on Advanced Robotics (ICAR)*; Ljubljana, Slovenia, 2021. pp. 338–43.

[28] Calinon S., Lee D. 'Learning control' in Vadakkepat P., Goswami A. (eds.). Humanoid robotics: a reference. Springer; 2019. pp. 1261–312.

[29] Robotics codes from scratch (RCFS). 2022. Available from https://robotics-codes-from-scratch.github.io

[30] Tanwani A.K., Yan A., Lee J., Calinon S., Goldberg K. 'Sequential robot imitation learning from observations'. *The International Journal of Robotics Research (IJRR)*. 2021, vol. 40(10–11), pp. 1306–25.

[31] Lembono T.S., Calinon S. 'Probabilistic iterative LQR for short time horizon MPC'. *2021 IEEE/RSJ International Conference on Intelligent Robots and Systems (IROS)*; Prague, Czech Republic, 2021. pp. 556–62.

[32] Girgin H., Jankowski J., Calinon S., Asfour T., Khatib O. 'Reactive anticipatory robot skills with memory'. *Proc. Intl Symp. on Robotic Research (ISRR)*; 2022.

[33] Paraschos A., Daniel C., Peters J, *et al.* 'Probabilistic movement primitives' in Burges C.J.C., Bottou L., Welling M. (eds.). Advances in neural information processing systems (neurips). USA: Curran Associates, Inc; 2013. pp. 2616–24.

[34] Jankowski J., Racca M., Calinon S. 'From key positions to optimal basis functions for probabilistic adaptive control'. *IEEE Robotics and Automation Letters*. 2022, vol. 7(2), pp. 3242–49.

[35] Bouguila N., Fan W. 'Mixture models and applications' in Cham; 2020.

[36] Ijspeert A.J., Nakanishi J., Hoffmann H., Pastor P., Schaal S. 'Dynamical movement primitives: learning attractor models for motor behaviors'. *Neural Computation*. 2013, vol. 25(2), pp. 328–73.

[37] Broomhead D.S., Lowe D. 'Multivariable functional interpolation and adaptive networks'. *Complex Systems*. 1988, vol. 2(3), pp. 321–55.

[38] Pennec X. 'Intrinsic statistics on riemannian manifolds: basic tools for geometric measurements'. *Journal of Mathematical Imaging and Vision*. 2006, vol. 25(1), pp. 127–54.

[39] Absil P.-A., Mahony R., Sepulchre R. Optimization algorithms on matrix manifolds. Princeton University Press; 2007.

[40] Park F.C., Bobrow J.E., Ploen S.R. 'A lie group formulation of robot dynamics'. *The International Journal of Robotics Research*. 1995, vol. 14(6), pp. 609–18.

[41] Selig J.M. Geometric fundamentals of robotics. Springer; 2005.

[42] Barfoot T.D., Furgale P.T. 'Associating uncertainty with three-dimensional poses for use in estimation problems'. *IEEE Transactions on Robotics*. 2014, vol. 30(3), pp. 679–93.

[43] Zeestraten M.J.A., Havoutis I., Silverio J., Calinon S., Caldwell D.G. 'An approach for imitation learning on riemannian manifolds'. *IEEE Robotics and Automation Letters (RA-L)*. 2017, vol. 2(3), pp. 1240–47.

[44] Jaquier N., Calinon S. 'Gaussian mixture regression on symmetric positive definite matrices manifolds: application to wrist motion estimation with semg'. *2017 IEEE/RSJ International Conference on Intelligent Robots and Systems (IROS)*; Vancouver, Canada, 2017. pp. 59–64.

[45] Lee T., Park F.C. 'A geometric algorithm for robust multibody inertial parameter identification'. *IEEE Robotics and Automation Letters*. 2018, vol. 3(3), pp. 2455–62.

[46] Traversaro S., Brossette S., Escande A., Nori F. 'Identification of fully physical consistent inertial parameters using optimization on manifolds'. *2016 IEEE/RSJ international conference on intelligent robots and systems (IROS)*; 2016. pp. 5446–51.

[47] Jaquier N., Rozo L., Caldwell D.G., Calinon S. 'Geometry-aware manipulability learning, tracking, and transfer'. *The International Journal of Robotics Research*. 2021, vol. 40(2–3), pp. 624–50.

[48] Park F.C., Kim J.W. 'Manipulability of closed kinematic chains'. *Journal of Mechanical Design*. 1998, vol. 120(4), pp. 542–48.

[49] Maurice P., Malaisé A., Amiot C., *et al.* 'Human movement and ergonomics: an industry-oriented dataset for collaborative robotics'. *The International Journal of Robotics Research.* 2019, vol. 38(14), pp. 1529–37.
[50] Jaquier N., Rozo L., Calinon S. 'Analysis and transfer of human movement manipulability in industry-like activities'. *2020 IEEE/RSJ International Conference on Intelligent Robots and Systems (IROS)*; 2020. pp. 11131–38.
[51] Yoshikawa T. 'Dynamic manipulability of robot manipulators'. *1985 IEEE International Conference on Robotics and Automation*; St. Louis, MO, USA, 1985. pp. 113–24.
[52] Dimeas F. 'Singularity avoidance in human-robot collaboration with performance constraints'. *Proc. Intl Workshop on Human-Friendly Robotics (HFR)*; 2020. pp. 93–98.
[53] Marić F., Petrović L., Guberina M., Kelly J., Petrović I. 'A riemannian metric for geometry-aware singularity avoidance by articulated robots'. *Robotics and Autonomous Systems*. 2021, vol. 145, p. 103865.
[54] Orin D.E., Goswami A. 'Centroidal momentum matrix of a humanoid robot: structure and properties'. *Proc. IEEE/RSJ Intl Conf. on Intelligent Robots and Systems (IROS)*; 2008. pp. 653–59.
[55] Jaquier N., Haschke R., Calinon S. 'Tensor-variate mixture of experts for proportional myographic control of a robotic hand'. *Robotics and Autonomous Systems*. 2021, vol. 142, 103812.
[56] Shetty S., Lembono T., Löw T, *et al.* 'Tensor train for global optimization problems in robotics'. *arXiv:220605077*. 2022.
[57] Mathew G., Mezic I. 'A uniform coveragealgorithm for mobile sensor networks'. *Proc. IEEE Conf. on Decision and Control*; 2009. pp. 7872–77.
[58] Ayvali E., Salman H., Choset H. 'Ergodic coverage in constrained environments using stochastic trajectory optimization'. *2017 IEEE/RSJ International Conference on Intelligent Robots and Systems (IROS)*; Vancouver, BC, 2017. pp. 5204–10.
[59] Abraham I., Prabhakar A., Murphey TD. 'An ergodic measure for active learning from equilibrium'. *IEEE Transactions on Automation Science and Engineering*. 2021, vol. 18(3), pp. 917–31.
[60] Shetty S., Silverio J., Calinon S. 'Ergodic exploration using tensor train: applications in insertion tasks'. *IEEE Transactions on Robotics*. 2021, vol. 38(2), pp. 906–21.

Chapter 3

A control framework for learning and performing safe human-robot collaborative assembly

Leonidas Koutras[1], Konstantinos Vlachos[1], George S. Kanakis[1], Fotios Dimeas[1], Zoe Doulgeri[1], and George A. Rovithakis[1]

The advent of collaborative robots has enabled new applications where human and robots can work together safely and effectively toward achieving a common goal. However, robots in the industry, and particularly in assembly lines, are still programmed with time-consuming methods, have little to no adaptability to changes in the environment, and still present safety issues in close collaboration with a human, due to stiff control methods utilized to achieve accuracy. In this chapter, we address the problem of learning a pick-and-place task by kinesthetic demonstration and executing it to new and even moving targets, aiming at collaborative assembly lines where operations take place on a moving conveyor. Moreover, the human safety aspect is addressed via a control framework that includes two layers. The first layer is in the form of collision avoidance by altering the executed trajectory, and the second one is achieved by a compliant robot behavior that preserves safety under unintentional contacts, without compromising target-reaching accuracy. For the learning part, we modify the traditional dynamic movement primitive (DMP) formulation for properly encoding the demonstrated task and generalizing it spatially to even moving targets. For the performance and safety part, we propose a control structure incorporating the DMP, the human avoidance signals, and the low stiffness controller of the robot to achieve safety and accuracy. Our approach is validated experimentally in a real-world industrial assembly scenario of a TV manufacturer, where a robot and a human collaboratively achieve the assembly of electronic boards on a TV back plate. This work contributes on a wider scale, as it aims to advance the way toward the adoption of human-robot collaboration in industrial assembly lines.

[1]Automation & Robotics Lab, Department of Electrical & Computer Engineering, Aristotle University of Thessaloniki, Thessaloniki, Greece

3.1 Introduction

Collaborative robots have opened up new ways for automating industrial processes in unstructured environments, where traditional industrial manipulators struggle. With smaller footprint and increased safety, they have become a part of Industry 4.0 and operate alongside humans in a shared workspace to accomplish tasks through collaboration [1–3]. Using teaching by demonstration, they can be programmed very quickly and can achieve human-likeness in their motion, which is an important aspect of user acceptance in close collaboration [4]. Encoding the demonstrated trajectory with parametric methods, the robot can autonomously execute the trajectory in a flexible and adaptable manner, surpassing simple record-and-play techniques. There have been many methods proposed in the literature that achieve these properties, such as dynamical systems [5], probabilistic methods [6], kernelized movement primitives [7], and dynamic movement primitives (DMPs) [8–10]. The popularity of DMP, in particular, is owed to their dynamical system structure, which allows seamless temporal and spatial scaling, as well as their augmentation with coupling terms for complex adaptive behaviors. A wide range of recent applications with DMP includes progressive automation [11], human-in-the-loop motion generation [12], adaptive trajectory generation under constraints [13], action recognition [14], and dual-arm manipulation [15].

Encoding complete trajectories in the task space, for example, with DMP, requires the representation of both the translational and the rotational components. The orientation is inherently more complicated than the position due to its manifold structure. One of the most popular orientation representations in the robotic literature is the unit quaternions [16]. Research on encoding quaternion trajectories with DMP has produced early results in References 17 and 18. These methods, however, could lead to oscillatory behaviors for some parameter values or demonstrated trajectories. In Reference 19, these issue is remedied with a new orientation DMP that is able to encode arbitrary quaternion trajectories.

A desired property of the method that is used to encode the demonstrated trajectory is the ability to generalize the learned pattern to new conditions. To this end, recent developments have equipped DMP with adaptive spatial scaling capabilities, for executing motions in dynamic environments. In Reference 20, a frame-independent spatial scaling method for DMP was proposed, to facilitate task-specific spatial scaling in the Cartesian space. In Reference 21, a DMP formulation for moving goals was proposed, in order to execute a pattern that has been demonstrated to a stationary goal. In this work, we combine these two methods to enable the successful generalization of trajectories in the case of an assembly operation that is taking place on a moving conveyor.

The safety aspect in human-robot collaboration is addressed at the hardware level by lightweight collaborative manipulator structures, with rounded edges and force and power limitations [22]. While this is a first step to avoid serious harm to the human, more advanced safety techniques should be utilized for the human-robot interaction scenarios to be both safe and effective in achieving a common collaborative task [23]. These techniques should prevent possible collisions by obstacle avoidance methods [24], as well as invoke appropriate reaction movements or introduce compliance in the manipulator upon detected collisions to prevent high contact forces [25, 26].

Robotic obstacle avoidance considering human-robot collaboration is an active research topic, with a focus on incorporating obstacle avoidance directly into the control architecture to ensure safety in dynamic environments [27–29]. Most obstacle avoidance methods can be integrated into DMP formulations through appropriate coupling terms. In Reference 30, an obstacle avoidance coupling term was introduced to displace the generated trajectory. This method, however, does not guarantee successful obstacle avoidance. In Reference 31, a potential field method was applied to a DMP setup that successfully avoids obstacles; however, it was only formulated for point obstacles. In Reference 32, this method was extended for obstacles enclosed in smooth surfaces. To successfully apply the aforementioned obstacle avoidance methods, the robot must accurately track this perturbed trajectory, which is a valid assumption for stiff position-controlled robots. In addition, the exact position of the obstacle needs to be provided by a perception system, so these methods are prone to perception accuracy, processing speed, and reliability. The majority of perception systems used in robotics cannot always guarantee these requirements. Perception errors or occlusions can raise safety issues. Another problem of obstacle avoidance techniques is in the acceleration and velocity bounds present in any real robot, which may not allow to steer the robot away from an obstacle. Although obstacle avoidance methods do have a solid mathematical foundation, collisions with the obstacles may still occur in practice due to the aforementioned limitations. Therefore, it is imperative to have a second level of safety, which can be in the form of active compliance on the entire robotic chain.

Compliant robotic behaviors have been developed in the literature in order to facilitate the safe interaction of the robot with its environment [33]. The most popular strategies for compliant control are impedance [34] and admittance [35], which transform the closed-loop robot dynamics into the form of a mass-spring-damper system. Thus, the system is able to interact with its environment, while remaining safe for both itself and humans, or other objects in its vicinity [36]. However, such methods are sensitive to unmodeled dynamics as well as task uncertainties. Such errors appear as disturbances to the system causing a drop in the tracking accuracy in the case of highly compliant target impedance. To remedy this issue, a control structure is proposed in our previous work [37] that utilizes a DMP with adaptive parameters and a robot under Cartesian impedance control with low apparent stiffness. With this structure, the robot remains compliant throughout the entire trajectory execution. Under the external force that arises in an unexpected contact, the generated trajectory is slowed down or even paused resulting in the robot behaving as a static low stiffness. After the external contact force disappears, the trajectory generation continues. This low-stiffness robot achieves accurate tracking of the generated trajectory, despite the disturbances created by modeling errors and unmodeled task or load dynamics, by modifying the desired trajectory according to an estimate of the modeling and task error bounds [37].

In this work, we present a learning and control framework for efficient human-robot collaboration, through easy teaching by demonstration, adaptable execution, and safe coexistence in a shared workspace. The framework integrates and extends our previous work in learning by the demonstration of the robot's position and orientation with DMP. It also integrates our work in control for accurate tracking performance of a compliant robot, to achieve accurate reaching of moving goals with two layers of human safety. These layers are based on obstacle avoidance and manipulator compliance under collisions. The methodology takes advantage of torque-controlled and redundant manipulators for full-body compliance and nullspace optimization, respectively. Our approach is validated experimentally in a real-world industrial assembly scenario of a TV manufacturer.

3.2 Learning by kinesthetic demonstration how to pick and place electronic boards on a TV

In this section, we initially present the introduction of human-robot collaboration in a TV assembly line, which exemplifies the application of our proposed framework. The operation is traditionally performed by two workers, one from the one side of the conveyor who picks the electronic boards from a container and places them on the back plate of the TV and another who fixes them with screws (Figure 3.1a). In general, electronic device manufacturing is a robot-willing industry that has not yet automated many of its assembly operations due to the uncertainties and complexity they entail. In the TV assembly cell, uncertainties arise in the location of the TV on the running conveyor that may produce board placement errors and impact the screwing operation. Fully automating this assembly with traditional programming is hence difficult without changing considerably the layout, introducing fixtures and other automation mechanisms to eliminate such uncertainties. Moreover, frequent changes in production (e.g., different TV models) require time-consuming reprogramming. This scenario is a good example to demonstrate how such industries can benefit from human-robot collaboration.

(a) (b)

Figure 3.1 The TV assembly cell as a use case for demonstrating the effectiveness of human-robot collaborative assembly

The envisioned collaborative cell consists of a robotic manipulator that is responsible for picking and placing the boards on the TV and a human worker who performs the screwing of the boards on the TV back plate, as shown in Figure 3.1b. The human and the robot share their workspace over the conveyor line in order to collaboratively assemble the two boards on the back plate. This task allocation between them minimizes the necessary changes in the production layout due to the robot introduction.

3.2.1 Encoding end-effector poses with dynamic movement primitives

The teaching of the pick-and-place task to the robot is performed by a human on a stationary conveyor. The human grabs the robot from anywhere and kinesthetically demonstrates how to move to pick each board, when to close the gripper, how to move to transfer it to the right place on the TV back plate, when to release the board by opening the gripper, and how to retract and repeat the process for the next board. Demonstrating the task into consecutive segments, by using a graphical interface such as a tablet or by pressing a button at the robot's wrist at the end of each segment, allows the association of gripper actions with certain motion segments.

Upon teaching completion, the demonstrated position and orientation trajectory of the robot's end-effector are encoded by DMP, and the robot can immediately start the autonomous execution. DMPs [8–10] form a compact representation to encode the demonstrated trajectories in the Cartesian space with a set of parameters that can be learned and generalized spatially and temporally. For encoding the TV assembly, the robot's trajectory for each segment is encoded in a separate DMP.

The DMP system for representing the position of the robot consists of a linear attractor and a nonlinear forcing term $\boldsymbol{F}_P(x)$ used to encode the desired trajectory shape:

$$\begin{aligned} \tau \dot{\boldsymbol{z}}_P &= \alpha_{z,P}(\beta_{z,P}(\boldsymbol{g} - \boldsymbol{y}_d) - \boldsymbol{z}_P) + \boldsymbol{S}(\boldsymbol{g} - \boldsymbol{y}_{0,d})\boldsymbol{F}_P(x) \\ \tau \dot{\boldsymbol{y}}_d &= \boldsymbol{z}_P \end{aligned} \tag{3.1}$$

where $\alpha_{z,P}$ and $\beta_{z,P}$ are positive gains, $\boldsymbol{y}_d, \boldsymbol{z}_P \in \mathbb{R}^3$ are the position and scaled velocity, $\tau > 0$ is a temporal scaling scalar, $\boldsymbol{g}, \boldsymbol{y}_{0,d} \in \mathbb{R}^3$ are the goal and initial position, $\boldsymbol{S}(\boldsymbol{g} - \boldsymbol{y}_{0,d}) \in \mathbb{R}^{3\times 3}$ is a spatial scaling matrix, and x is the phase variable, utilized as a clocking mechanism to avoid explicit time dependency. The dynamics of the phase variable are governed by the canonical system:

$$\tau \dot{x} = -\alpha_x x \tag{3.2}$$

with its initial value $x_0 = 1$ and α_x a positive gain.

The forcing term $\boldsymbol{F}_P(x)$ is instrumental in the encoding of the desired trajectory and is given by

$$\boldsymbol{F}_P(x) = \frac{\sum_{i=1}^{N} \boldsymbol{w}_i \Psi_i(x)}{\sum_{i=1}^{N} \Psi_i(x)} x \tag{3.3}$$

where $\Psi_i(x) = \exp(-h_i(c_i - x)^2)$ are N Gaussian kernel functions, with h_i and c_i denoting their inverse widths and centers, respectively. The weights $\boldsymbol{w}_i \in \mathbb{R}^3$ are learned using locally weighted regression to encode the desired trajectory.

The position DMP presented in (3.1) does not cater for position targets that are time-varying, because in this case, the linear dynamics part is no longer a linear attractor to the moving goal. A solution for this problem is provided in the next section.

To encode orientation trajectories, the majority of the literature utilizes unit quaternions [16]. Early works in encoding orientation trajectories with DMP can be found in References 17 and 18. However, these approaches can produce undesired oscillations in the robot's orientation, producing a motion pattern highly deviant from the desired. A correct formulation proposed in Reference 19 alleviates such problems while guaranteeing the generation of orientation parameters that lie in the unit quaternion space. The transformation system of this corrected formulation given $\boldsymbol{Q}_0, \boldsymbol{Q}_g$ the initial and goal orientation expressed as unit quaternions is given by

$$\begin{aligned} \tau \dot{\boldsymbol{z}}_Q &= -\alpha_{z,Q}(\beta_{z,Q}\boldsymbol{e}_Q + \boldsymbol{z}_Q) + \boldsymbol{S}(\boldsymbol{e}_{Q,0})\boldsymbol{F}_Q(x) \\ \tau \dot{\boldsymbol{e}}_Q &= \boldsymbol{z}_Q \end{aligned} \tag{3.4}$$

where $\boldsymbol{e}_Q = 2\log \boldsymbol{Q}_g * \overline{\boldsymbol{Q}}_d$ is the orientation error from the goal $\boldsymbol{Q}_g$ with $\boldsymbol{e}_{Q,0} = 2\log \boldsymbol{Q}_g * \overline{\boldsymbol{Q}}_0$ its initial value where the dashed quaternions denote their conjugate or inverse, $\boldsymbol{z}_Q$ its velocity scaled by τ, and $\alpha_{z,Q}$ and $\beta_{z,Q}$ are positive gains. The reference orientation is calculated from the generated $\boldsymbol{e}_Q$:

$$\boldsymbol{Q}_d = \begin{bmatrix} \eta & \boldsymbol{\varepsilon}^T \end{bmatrix}^T = \exp\left(-\frac{1}{2}\boldsymbol{e}_Q\right) * \boldsymbol{Q}_g \tag{3.5}$$

and the reference angular velocity is calculated from the orientation error derivative $\dot{\boldsymbol{e}}_Q$ using the appropriate Jacobian matrix [19, 38]. The forcing term $\boldsymbol{F}_Q(x)$ is calculated in accordance with that of the position DMP from (3.3), and the canonical system is common with the position (3.2). The scaling matrix $\boldsymbol{S}(\boldsymbol{e}_{Q,0})$ is defined similarly to the position DMP, as explained in section 3.2.2.

3.2.2 Frame-independent spatial scaling of learned trajectories

In the original DMP formulation, the spatial scaling matrix is given as $\boldsymbol{S}(\boldsymbol{g} - \boldsymbol{y}_{0,d}) = \text{diag}(\boldsymbol{g} - \boldsymbol{y}_{0,d})$, with diag(·) denoting a diagonal matrix with elements of the components of its argument vector. However, as explained in References 20 and 30, this scaling can be problematic. The problems stem from the fact that scaling is applied separately to each Cartesian coordinate, and therefore geometric characteristics of the shape of the path may be lost. One can better understand the consequences of this problem by imagining a task where in one coordinate the initial and goal position are the same. Then the corresponding element of the scaling matrix would be zero, resulting in no motion in this direction. To remedy this problem, a DMP scaling was proposed in Reference 20 that scales the trajectory by a constant and rotates it by

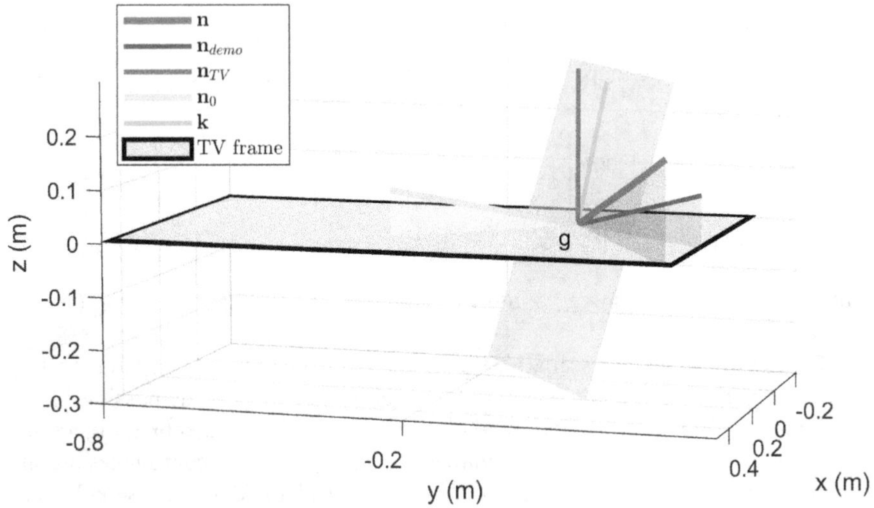

Figure 3.2 *Visualization of the frame-independent spatial scaling around axis* $\boldsymbol{k}$*, which is the projection of the normal to the TV* $\boldsymbol{n}_{TV}$ *on the green plane. The green plane is normal to the vector* $\boldsymbol{n}_0$.

multiplying it with the appropriate rotation matrix that depends on the task. This task dependency stems from the selection of the axis about which the trajectory is rotated. In particular, $\boldsymbol{S}(\boldsymbol{g}-\boldsymbol{y}_{0,d}) = s_g\boldsymbol{R}_g$, where $s_g = \frac{\|\boldsymbol{g}-\boldsymbol{y}_{0,d}\|}{\|\boldsymbol{g}_{\text{demo}}-\boldsymbol{y}_{0,\text{demo}}\|}$ with $\boldsymbol{g}_{\text{demo}}$ and $\boldsymbol{y}_{0,\text{demo}}$ denoting the goal and initial position of the demonstrated trajectory is a magnitude scaling constant and $\boldsymbol{R}_g \in SO(3)$ a rotation matrix that rotates $\boldsymbol{n}_{\text{demo}} = \frac{\boldsymbol{g}_{\text{demo}}-\boldsymbol{y}_{0,\text{demo}}}{\|\boldsymbol{g}_{\text{demo}}-\boldsymbol{y}_{0,\text{demo}}\|}$ to $\boldsymbol{n} = \frac{\boldsymbol{g}-\boldsymbol{y}_{0,d}}{\|\boldsymbol{g}-\boldsymbol{y}_{0,d}\|}$ about an appropriately selected axis.

The selection of the rotation matrix is not unique, however. In fact, one can calculate the space of valid rotation axes as follows. Let $\boldsymbol{n}_{\text{demo}}$ and $\boldsymbol{n}$, visualized with blue and red vectors in Figure 3.2, be put on an arbitrary common starting point. The set of available rotation axes belongs on the plane dissecting the angle between the two vectors visualized with green, which is determined by its normal $\boldsymbol{n}_0 = \frac{\boldsymbol{n}-\boldsymbol{n}_{\text{demo}}}{\|\boldsymbol{n}-\boldsymbol{n}_{\text{demo}}\|}$, visualized with cyan. All unit vectors on this plane constitute a valid rotation axis for $\boldsymbol{R}_g$. Thus, the set of candidate rotation axes can be expressed as follows:

$$S_k = \{\boldsymbol{k} \in \mathbb{R}^3 \mid \boldsymbol{k}^T\boldsymbol{n}_0 = 0, \|\boldsymbol{k}\| = 1\} \tag{3.6}$$

After selecting a suitable rotation axis, the subsequent rotation angle is given by

$$\theta = \cos^{-1}\left(\frac{\boldsymbol{n}_{\text{demo}}^T(\boldsymbol{I}_3 - \boldsymbol{k}\boldsymbol{k}^T)\boldsymbol{n}}{\|(\boldsymbol{I}_3 - \boldsymbol{k}\boldsymbol{k}^T)\boldsymbol{n}_{\text{demo}}\|\,\|(\boldsymbol{I}_3 - \boldsymbol{k}\boldsymbol{k}^T)\boldsymbol{n}\|}\right) \tag{3.7}$$

Then, rotation matrix $\boldsymbol{R}_g$ is computed through Rodrigues' rotation formula [39]:

$$R_g = I_3 + \hat{k}\sin\theta + \hat{k}^2(1 - \cos\theta) \tag{3.8}$$

where the hat operator maps a three-dimensional (3D) vector to a skew-symmetric 3×3 matrix:

$$\hat{v} = \begin{bmatrix} 0 & -v_3 & v_2 \\ v_3 & 0 & -v_1 \\ -v_2 & v_1 & 0 \end{bmatrix}$$

Notice that by substituting $n_0 = n - n_d = (R_g - I_3)n_d$ and (3.8) to $k^T n_0$, (3.6) is satisfied for all unit k.

For the TV assembly scenario, the rotation R_g utilized in the spatial scaling of the learned trajectory is selected to be about an axis as close as possible to the normal of the TV that is denoted by n_{TV} and visualized with magenta in Figure 3.2. Such a selection of rotation axis minimizes trajectory rotation coordinates around the two axes other than n_{TV}. Such rotations, if considerable, can cause collisions with the TV board, and thus the probability of generating trajectories that lead to collisions is minimized. Following the analysis in Reference 20, the selected rotation axis k should be the projection of n_{TV} to the space of available rotation axes S_k. Therefore, the selected rotation axis is given by

$$k = \frac{(I_3 - n_0 n_0^T)n_{TV}}{\|(I_3 - n_0 n_0^T)n_{TV}\|} \tag{3.9}$$

The rotation axis is visualized in Figure 3.2 with a green vector on the plane.

3.3 Adaptable DMP execution to a moving goal

After the demonstration is completed and the task has been encoded, the DMP can be used to generate reference trajectories for the robot that reproduce the learned motion pattern and autonomously perform the picking and placing of electronic boards. However, goal locations will be in general different from those of the demonstration, so the proposed scaling needs to be applied. The location of the boards in the container is well defined and equidistant, for the robot to easily pick them, simply by modifying the target position of the respective DMP segment. On the other hand, the board placement location on the TV is dependent on the TV back plate pose on the conveyor, which is not fixed and is moreover constantly moving. As a result, the DMP should scale the learned trajectory segments to a moving goal.

Traditional DMP methods are formulated with stationary goals in mind and are therefore unable to successfully generalize to a moving goal, as in this TV assembly scenario. In that case, the phase variable may vanish before completing the desired trajectory and therefore generalize the learned pattern incorrectly. Furthermore, once the target is reached, the traditional DMP does not guarantee accurate trajectory tracking. The DMP execution toward a moving target on the conveyor belt needs to adapt the trajectory online, according to the moving goal, and track the target after the DMP has reached it. The problem of DMP with moving goals has been studied in Reference 21, where the source of this problem was identified to be the structure of the transformation system. While a stationary goal is a stable equilibrium of the linear part of the transformation, the system with a moving goal will not converge to a stable equilibrium of zero tracking error. The following transformation system remedies this problem and hence is suitable for use with moving targets:

$$\begin{aligned} \tau\dot{z}_P &= -\alpha_{z,P}(\beta_{z,P}\boldsymbol{e}_{P,G} + z_P) + \boldsymbol{S}(\boldsymbol{e}_{P,0})\boldsymbol{F}_P(x) \\ \tau\dot{\boldsymbol{e}}_{P,G} &= z_P \end{aligned} \tag{3.10}$$

where $\boldsymbol{e}_{P,G} = \boldsymbol{g} - \boldsymbol{y}_d$, the position error from the target, and $\boldsymbol{e}_{P,0} = \boldsymbol{g} - \boldsymbol{y}_{0,d}$, its initial value. With z_P, we denote the scaled by τ velocity error. The gains and temporal scaling are defined as in (3.1). Spatial scaling is defined as explained in section 3.2.2.

The formulation of (3.10) can indeed handle moving goals by utilizing the tracking error dynamics for the transformation system linear part, instead of the position dynamics of (3.1). By formulating the DMP anchored at the goal rather than on the inertia frame, i.e., with respect to the error from the goal, the resulting trajectory is much better generalized during execution to a moving goal. Furthermore, since a zero tracking error is a stable equilibrium of the linear part of (3.10), the goal trajectory is accurately tracked after the reaching pattern is completed. To obtain the desired trajectory $\boldsymbol{y}_d$, measurements of the goal's position and velocity are required by this formulation in order to get $\boldsymbol{y}_d = \boldsymbol{g} - \boldsymbol{e}_{P,G}$ and $\dot{\boldsymbol{y}}_d = \dot{\boldsymbol{g}} - \frac{1}{\tau}z_P$. These measurements are provided in our case by the visual perception system, which tracks the moving TV.

In the training phase, a stationary TV frame is used. Notice that, for a stationary goal, substituting $\boldsymbol{e}_{P,G} = \boldsymbol{g} - \boldsymbol{y}_d$ and $\dot{\boldsymbol{e}}_{P,G} = \dot{\boldsymbol{g}} - \dot{\boldsymbol{y}}_d = -\dot{\boldsymbol{y}}_d$ to (3.10) yields (3.1) but with the forcing term negated. Thus, when training the DMP with (3.1), execution with (3.10) requires the use of $-\boldsymbol{F}_P(x)$ in place of the desired forcing term. Alternatively, one could directly utilize (3.10) for training.

In Reference 21, an additional equation for online adaptation of the temporal scaling parameter was proposed. This equation allows trajectory generation with a similar velocity profile as in the demonstration. Therefore, for goals moving further away than the goal in the demonstration, the trajectory execution would be slower. However, in time-critical applications such as ours, this is not desired, as the robot must quickly place the boards on the TV and retract, so that the worker has enough time to screw them.

3.4 A robot control framework to ensure safe and effective human-robot collaboration

The basic idea of the proposed approach is to add two layers of safety because the robot will share its workspace with the human. The first layer involves avoidance of the human, which can be considered as a dynamic obstacle, by modifying the DMP trajectory in the task space as well as by inducing additional motion in the nullspace of the robot in order to fully exploit the redundant degrees of freedom.

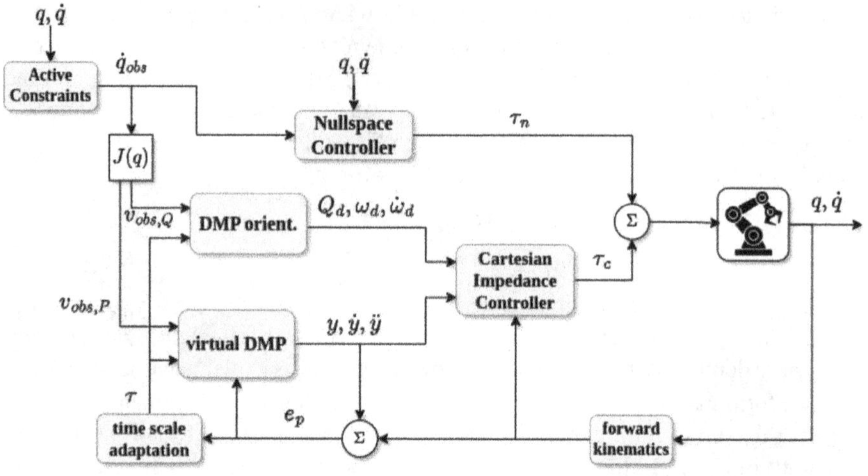

Figure 3.3 Architecture of the proposed control scheme

The second level of safety is achieved by making the robot compliant throughout the autonomous execution, in order to be safe in case a collision is unavoidable. However, compliance needs to be achieved without compromising accuracy. To this aim, we propose a framework that is based on a low stiffness impedance controlled robot driven by a control structure that achieves high accuracy in the task space for reaching the desired target.

The proposed solution is shown as a block diagram in Figure 3.3. A low-stiffness impedance robot controller in the Cartesian space is provided by a reference position and orientation trajectory generated by a virtual DMP for the position and an orientation DMP both augmented with coupling terms arising from an obstacle avoidance signal added at the velocity level. The virtual position DMP is the trained DMP modified appropriately for accurately tracking the desired position trajectory under modeling and task errors and allowing highly compliant reactions at unexpected collisions.

3.4.1 Combining compliance, tracking accuracy, and obstacle avoidance

To achieve compliant robot behavior and at the same time high tracking accuracy in following the desired position trajectory and reaching of the target, we utilize a virtual DMP system initially proposed in our previous work [37]. This virtual DMP incorporates a position tracking error feedback and produces a position reference trajectory $\boldsymbol{y}, \dot{\boldsymbol{y}}, \ddot{\boldsymbol{y}} \in \mathbb{R}^3$ which is able to compensate for errors due to model and task uncertainties. The virtual DMP is augmented by an obstacle avoidance coupling term at the velocity level $\boldsymbol{v}_{\text{obs},P}$, which further modifies the reference position trajectory to account for obstacles existing in the task space of the robot. The virtual DMP transformation system is given by

$$\tau \dot{z}_P = -a_z(\beta_z e_{P,G} + z_P - e_p) + S(e_{P,0}) F_P(x)$$
$$\tau \dot{e}_{P,G} = z_P - v_{\text{obs},P} \tag{3.11}$$

where $e_p = p - y \in \mathbb{R}^3$ is the position tracking error with $p \in \mathbb{R}^3$ being the end-effector position acquired via the forward robot kinematics from measured joint positions. Notice that if $e_p = 0$ the reference trajectory $y, \dot{y}, \ddot{y}$ produced by the virtual DMP is the desired trajectory described in terms of y_d, $\dot{y}_d$, $\ddot{y}_d$, which owes itself to the application of the forcing term $F_P(x)$ to the virtual DMP system. In case $e_p \neq 0$, the reference trajectory is displaced from the desired by the amount needed so that the robot end-effector position p accurately tracks the desired trajectory y_d. Regarding the orientation, the reference trajectory in terms of $Q_d, \omega_d, \dot{\omega}_d$ is produced by the orientation DMP system (3.4) augmented by $v_{\text{obs},Q}$, which is the respective orientation obstacle avoidance coupling term:

$$\tau \dot{z}_Q = -\alpha_z(\beta_z e_Q + z_Q) + S(e_{Q,0}) F_Q(x)$$
$$\tau \dot{e}_Q = z_Q - v_{\text{obs},Q} \tag{3.12}$$

The minus sign in the last equation of (3.12) owes itself to the integration of the orientation error e_Q. Notice that in contrast with the virtual DMP, the orientation DMP does not utilize feedback of the orientation tracking error, and thus the reference orientation trajectory does not compensate for errors due to model and task uncertainties. Both systems (3.11) and (3.12) utilize a common canonical system which is adapting the time evolution of the DMP via a sigmoid time scale adaptation law initially presented in Reference 37 in order to remain synchronized as well as enable the robot to be compliant to external contact forces arising in case of unexpected collision. In particular, the sigmoid time scale adaptation is given by

$$\tau = \tau_0 \left(1 + \exp\left(a_{\text{sig}}\left(\left\|e_p\right\| - c_{\text{sig}}\right)\right)\right) \tag{3.13}$$

with $\tau_0 > 0$ being the demonstrated duration of the movement and $a_{\text{sig}}, c_{\text{sig}} \in \mathbb{R}^+$ being positive parameters of the sigmoid function that regulate its slope and its center. This specific implementation of the time scaling adaptation enables the abrupt increase in τ when the norm of the position tracking error e_p exceeds a certain value stopping the evolution of the system (3.11) and (3.12). Thus, the virtual DMP provides a constant position input as a reference to the compliant robot control system, resulting in the reduction of the whole control structure into an impedance-controlled robot driven by external forces, such as the regulation case.

The robot control system realizes target inertia, damping, and stiffness M_d, D_d, K_d (positive definite diagonal matrices) with respect to the position and the orientation tracking error $\begin{bmatrix} e_p^T & e_o^T \end{bmatrix}^T$ expressed by the following closed-loop dynamics:

$$M_d \begin{bmatrix} \ddot{e}_p \\ \dot{\omega}_e \end{bmatrix} + D_d \begin{bmatrix} \dot{e}_p \\ \omega_e \end{bmatrix} + K_d \begin{bmatrix} e_p \\ e_o \end{bmatrix} = F_e + F_{\text{ext}} \tag{3.14}$$

where $e_o = 2\log(Q_e) \in \mathbb{R}^3$ is the orientation tracking error with $Q_e = Q * \overline{Q}_d$ being the unit quaternion tracking error, Q the end-effector orientation, Q_d the desired orientation provided by the orientation DMP, $\boldsymbol{\omega}_e = \boldsymbol{\omega} - \boldsymbol{\omega}_R \in \mathbb{R}^3$ is the angular velocity error with $\boldsymbol{\omega} \in \mathbb{R}^3$ being the angular velocity of the end-effector, $\boldsymbol{\omega}_R = \boldsymbol{R}_e\boldsymbol{\omega}_d \in \mathbb{R}^3$ is the geometrically consistent desired angular velocity [38, 40], with $\boldsymbol{\omega}_d \in \mathbb{R}^3$ being the reference angular velocity provided by the orientation DMP and with $\boldsymbol{R}_e \in \mathbf{SO}(3)$ denoting the rotation matrix corresponding to $\boldsymbol{Q}_e$, and $\boldsymbol{F}_e(t) \in \mathbb{R}^6$ is a disturbance term stemming from unmodeled dynamics like unmodeled joint friction, residual dynamics due to uncertain robot dynamic parameters, as well as task uncertainties (e.g., load) and external force measurement inaccuracies. Finally, $\boldsymbol{F}_{\text{ext}} \in \mathbb{R}^6$ are the external contact forces that may arise from unexpected collisions.

In case of a motion in the free space ($\boldsymbol{F}_{\text{ext}} = \mathbf{0}$) and considering the ideal case where the disturbance term $\boldsymbol{F}_e(t)$ is zero, the robot is able to track perfectly the reference trajectory. However, since $\boldsymbol{F}_e(t)$ exists in practice, the tracking accuracy of the reference trajectory in both position and orientation depends on the target impedance parameter values. To ensure safety under unintentional contact, we select low stiffness and damping gains for the position part and account for tracking accuracy via the reference trajectory provided by the virtual DMP. For the orientation part, we select high stiffness and damping gains to reduce the effect of the disturbance input to the orientation variables and thus achieve high orientation tracking accuracy as the impact of these variables on safety is not significant.

3.4.2 *Obstacle avoidance in operational and null spaces*

To achieve obstacle avoidance, we impose active constraints on the control system by generating a joint velocity signal $\dot{\boldsymbol{q}}_{\text{obs}} \in \mathbb{R}^n$. This velocity is decomposed into two components, one affecting the task space and another the robot nullspace. The nullspace component of the obstacle avoidance velocity is applied directly at the torque level utilizing a decoupled nullspace controller inspired by References 34 and 41, while the task space component is provided as a coupling term to the virtual DMP system (3.11), (3.12). The coupling term can be derived by

$$\boldsymbol{v}_{\mathbf{obs}} = \begin{bmatrix} \boldsymbol{v}_{\text{obs,P}} \\ \boldsymbol{v}_{\text{obs,Q}} \end{bmatrix} = \boldsymbol{J}(\boldsymbol{q})\dot{\boldsymbol{q}}_{\text{obs}} \tag{3.15}$$

where $\boldsymbol{J}(\boldsymbol{q})$ is the robot Jacobian.

The torque applied by the decoupled nullspace controller augments the impedance torque input that imposes the dynamics of (3.14) and is given by

$$\boldsymbol{\tau}_n = -\boldsymbol{N}^T\boldsymbol{D}_N\boldsymbol{N}(\dot{\boldsymbol{q}} - \dot{\boldsymbol{q}}_{\text{obs}}) \tag{3.16}$$

where $\boldsymbol{q}, \dot{\boldsymbol{q}} \in \mathbb{R}^n$ are the joint positions and velocities of the robot, $\boldsymbol{D}_N \in \mathbb{R}^{r\times r}$ is a diagonal damping matrix, and $\boldsymbol{N}(q) \in \mathbb{R}^{r\times n}$ is the dynamically consistent nullspace projection matrix, with n being the degrees of freedom of the robot and r being the redundant degrees of freedom. $\boldsymbol{N}(q)$ is given by

$$\boldsymbol{N}(\boldsymbol{q}) = \left(\boldsymbol{Z}(\boldsymbol{q})\boldsymbol{M}(\boldsymbol{q})\boldsymbol{Z}^T(\boldsymbol{q})\right)^{-1}\boldsymbol{Z}(\boldsymbol{q})\boldsymbol{M}(\boldsymbol{q}) \tag{3.17}$$

where $\boldsymbol{M}(\boldsymbol{q}) \in \mathbb{R}^{n \times n}$ is the inertia matrix of the robot and $\boldsymbol{Z}(\boldsymbol{q}) \in \mathbb{R}^{r \times n}$ is the nullspace base matrix, given by

$$\boldsymbol{Z}(\boldsymbol{q}) = \begin{bmatrix} -\boldsymbol{J}_r^T(\boldsymbol{q})\boldsymbol{J}_m^{-T}(\boldsymbol{q}) & \boldsymbol{I}_r \end{bmatrix} \tag{3.18}$$

where $\boldsymbol{I}_r$ is the identity matrix of dimension r, and matrices $\boldsymbol{J}_m$ and $\boldsymbol{J}_r$ can be computed by writing the Jacobian of the manipulator as

$$\boldsymbol{J}(\boldsymbol{q}) = \begin{bmatrix} \boldsymbol{J}_m(\boldsymbol{q}) & \boldsymbol{J}_r(\boldsymbol{q}) \end{bmatrix} \tag{3.19}$$

where $\boldsymbol{J}_m(\boldsymbol{q}) \in \mathbb{R}^{m \times m}$ is invertible away from singularities. Notice that the form of (3.19) may require the reordering of the columns of $\boldsymbol{J}(\boldsymbol{q})$ and that the nullspace matrix satisfies the equality $\boldsymbol{J}(\boldsymbol{q})\boldsymbol{Z}^T = \boldsymbol{0}$.

The nullspace part of the obstacle avoidance signal does not interfere with the execution of the task in the task space of the end-effector, and thus it does not induce errors that could cause the increase of the temporal scaling parameter and stop the evolution of the DMP.

3.4.3 Generation of obstacle avoidance velocity signal

A joint velocity signal $\dot{\boldsymbol{q}}_{\text{obs}}$ for obstacle avoidance is produced utilizing an extension of the method proposed in Reference 27 to account for the whole body of a robotic manipulator, considering multiple moving obstacles. This method enforces dynamic active constraints on moving obstacles modeled as convex enclosing surfaces. The method was originally proposed to avoid obstacles in the end-effector workspace of the robot by creating a velocity signal in the Cartesian space that guarantees collision avoidance of the end-effector of the robot with moving obstacles. The stability of the controller as well as the enforcement of the constraints was rigorously proven. Consider a single obstacle in the end-effector workspace. This velocity signal is produced using:

$$\boldsymbol{v} = -\kappa \frac{\varepsilon}{1+\xi} \frac{1}{\|\boldsymbol{v}_c\|^2} \boldsymbol{v}_c \tag{3.20}$$

with κ being a positive gain, $\xi = \frac{d(t)-\rho}{\rho}$, $\varepsilon := T(\xi) = \ln(1+\xi)$, ρ being a design parameter, and $\boldsymbol{v}_c = \left(\frac{\partial d(t)}{\partial \boldsymbol{p}}\right)^T$. The Euclidean distance from the end-effector of the robot is denoted with $d(t)$ with $\boldsymbol{p}$ being the actual position of the end-effector. This control input via the transformation of the error utilizing the diffeomorphism $T(\xi)$ guarantees that ε is bounded from below, thus ensuring that $d(t) > 0$ for all $t > 0$. For avoiding multiple obstacles, a combining function of the individual distances is utilized.

To generalize Reference 27 to the full body of the robot, we must initially model it. Using the approach explained in Reference 42, the whole body of the robot is covered with smooth enclosing surfaces. Specifically, each link is covered by a single capsule, which consists of two spheres, with centers connected over the axis of a cylinder. The sphere centers are placed on two consecutive joints, and the radii of the spheres and the cylinder are the same. Notice that by utilizing capsules, the distance

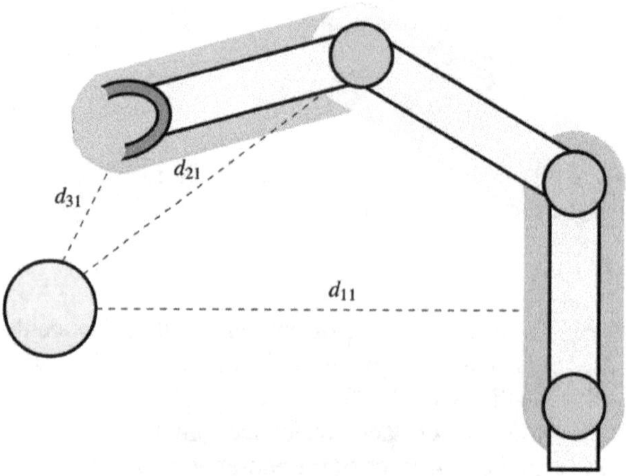

Figure 3.4 *The minimum distances of a spherical object (obstacle) from the capsules that cover the body of an example manipulator*

between any point and the boundary of the capsule can be computed analytically. The coverage of a three-degree-of-freedom manipulator by capsules is shown in Figure 3.4, as well as the distances from an obstacle modeled as a sphere.

The obstacle avoidance signal can be generated using a metric that combines the distances from each capsule to each obstacle. This metric must be positive when the robot does not collide with any obstacle and zero when an obstacle touches the boundary of the union of all capsules. For this work, we chose the harmonic mean, which is defined for $x_1, x_2, \ldots, x_n$, with $x_i \geq 0$ as

$$h(x_1, x_2, \ldots, x_n) = \frac{n}{\sum_{i=1}^{n} \frac{1}{x_i}} \tag{3.21}$$

We can also define the weighted harmonic mean as

$$h_w(x_1, x_2, \ldots, x_n) = \frac{\sum_{i=1}^{n} w_i}{\sum_{i=1}^{n} \frac{w_i}{x_i}} \tag{3.22}$$

Assuming that M capsules are needed in order to cover the full body of the robot and that N obstacles are in its workspace, we denote Euclidean distance $d_{ij}(t)$ between capsule i and obstacle j. We can further define the utilized metric function used for the obstacle avoidance signal as

$$m(t) = h(h_{w1}, h_{w2}, \ldots, h_{wn}) \tag{3.23}$$

where $h_{wi} = h_w(d_{i1}(t), d_{i2}(t), \ldots, d_{in}(t))$.

As $m(t)$ is the weighted harmonic mean for each capsule of the individual harmonic means for each obstacle, if one of the distances is zero, $m(t)$ becomes zero. Thus, $m(t) > 0, \forall t$ is a sufficient condition for successful obstacle avoidance. Furthermore, it demonstrates favorable properties, as it emphasizes more on small distances, making obstacles in the vicinity of the robot cause larger movements, without, however, completely disregarding distant obstacles. The selection of the weighted harmonic mean provides the possibility for task-specific prioritization of the motion of each capsule.

The final obstacle avoidance joint velocity used in (3.15) and (3.16) has the form of (3.15) and is given by

$$\dot{\boldsymbol{q}}_{\text{obs}} = -\kappa \frac{\varepsilon}{1+\xi} \frac{1}{\|\dot{\boldsymbol{q}}_c\|^2} \dot{\boldsymbol{q}}_c \tag{3.24}$$

where $\xi = \frac{m(t)-\rho}{\rho}$ and $\dot{\boldsymbol{q}}_c = \left(\frac{\partial m(t)}{\partial q}\right)^T$. All other parameters are the same as in (3.15). This control input guarantees again that ε is bounded from below for all $t > 0$ and thus $m(t) > 0$, ensuring collision avoidance of the full body of the robot with moving obstacles.

3.5 Experimental validation

To validate the proposed framework, we performed a series of experiments on a TV assembly line mockup that is shown in Figure 3.5. It consists of a KUKA LBR iiwa 7 robot, two RGBD cameras, a conveyor line, and a container for the electronics boards. The robot is in torque control mode, and we have implemented the desired target impedance without inertia reshaping. The proposed control framework is

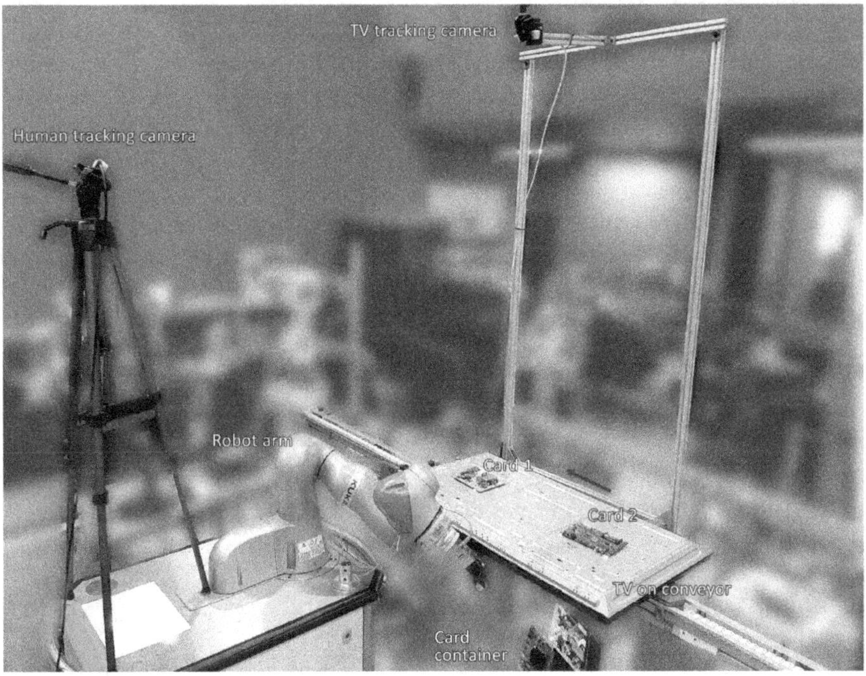

Figure 3.5 Experimental setup of a TV assembly cell mock-up with a conveyor

Table 3.1 System parameters in experiments

Parameter	Description	Value
α_z, β_z	DMP gains	30, 7.5
N	Number of kernels	80
diag($\boldsymbol{K}_d$)	Stiffness in position and orientation	200 N/m, 100 Nm/rad
diag($\boldsymbol{D}_d$)	Damping in position and orientation	35 Ns/m, 20 Nms/rad
$c_{\text{sig}}, \alpha_{\text{sig}}$	Sigmoid center and slope	0.07, 100
d_N	Nullspace gain	1
κ, ρ	Obstacle avoidance parameters	8, 0.38

running on an external computer with Ubuntu and communicates with the robot using the KUKA FRI library at 3 ms cycle time. The RGBD camera mounted above the conveyor is used to detect and track the moving TV, based on PVNet pose estimation [43, 44]. The other RGBD camera is used for obstacle avoidance and tracks the human, using an open-source skeleton tracking [45]. The control system parameters used in the experiments for the DMP encoding and for the autonomous execution are given in Table 3.1.

3.5.1 Teaching by demonstration

For the teaching phase, the robot is in gravity compensation mode with zero stiffness and damping. Before the teaching starts, the TV is placed on the stationary conveyor, and its pose is determined by the perception system. This is necessary to calculate the relative pose of each board with respect to the TV frame. Then, the operator grabs the robot and demonstrates the pick and placement of the first board (yellow one) in five segments, as shown in Figure 3.6a-e. The process is repeated for the second board (green one) because it has different dimension (Figure 3.6f). The keyboard on the computer is utilized by another human to log the start and end of each segment. Alternatively, utilizing the buttons on the robot's wrist, the human who is doing the demonstration could indicate the segments. After each segment is demonstrated, the DMPs are encoded and saved on the computer. The teaching is completed when all the predefined segments are demonstrated (five for each board). The discrete segments of the learned task, as they have been encoded with DMP, are depicted in Figure 3.7. The duration to teach how to pick and place both boards is 41 seconds. The ability to encode the pick and place task with a single demonstration in such a short period indicates the benefit of this approach over conventional robot programming.

3.5.2 Autonomous execution with a moving conveyor

After the teaching is complete, the robot can execute the demonstrated task autonomously and collaborate with the human to complete the assembly. A TV is placed on the conveyor that is moving approximately at 3 cm/s, equal to the speed of the

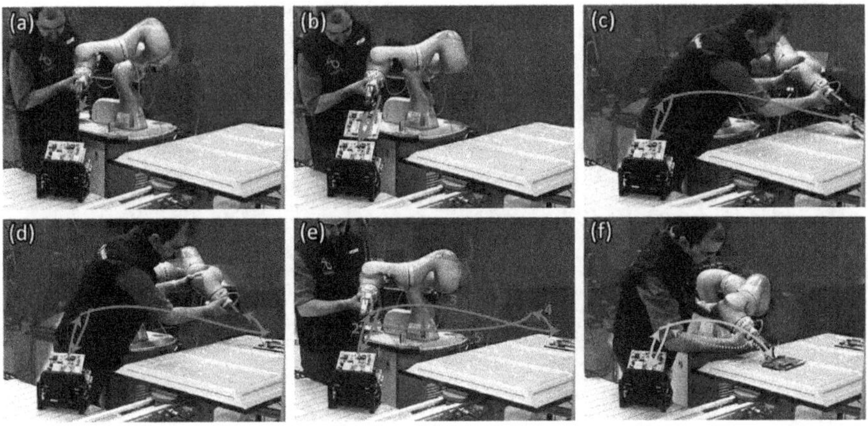

Figure 3.6 *Teaching by demonstration of the two boards. The operator initially picks and places the first board (a–f) by demonstrating five segments (red lines). Then repeats the teaching for the second board (green lines).*

actual production line. The pose of the TV with respect to the robot is tracked by the perception system, and the actual velocity of the TV frame is estimated online.

An illustration of the robot's position, orientation, and phase variable for all segments during the autonomous pick and place of the two boards is shown in Figure 3.8. For each segment, the targets to the DMP for picking the boards

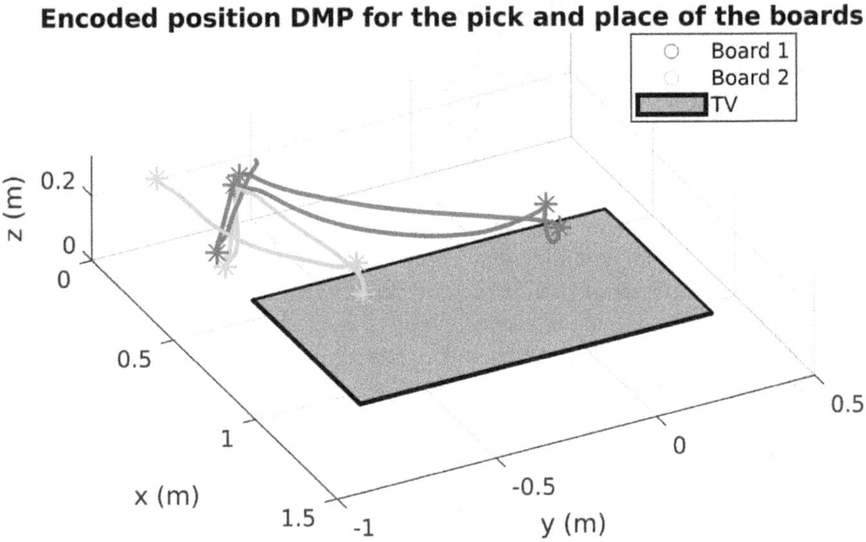

Figure 3.7 *Encoded translational components with DMP for each board. The asterisks represent the segments.*

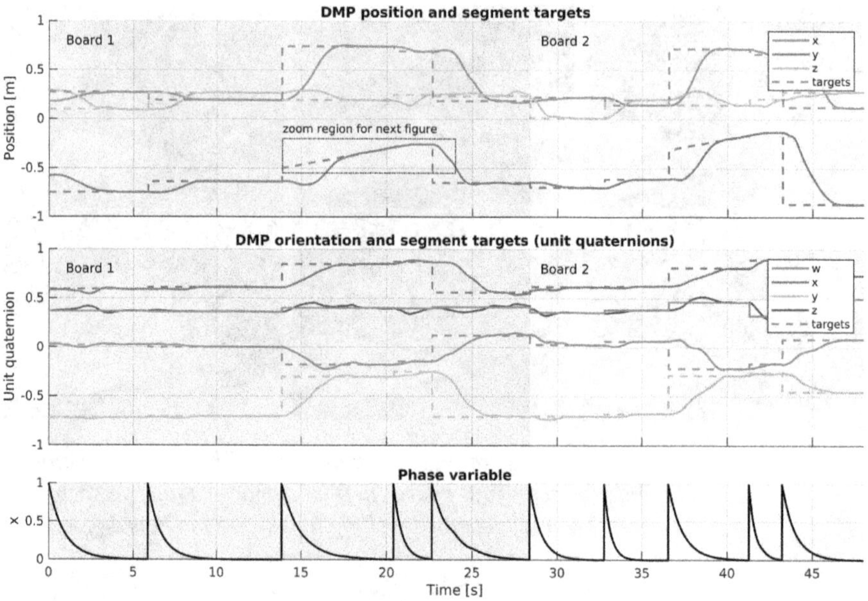

Figure 3.8 *Adaptable execution of the DMP, while the TV is moving on the conveyor. Top: the translational component of the DMP and the corresponding targets of each segment. Middle: the rotational component represented as unit quaternions. Bottom: the phase variable, indicating the different segments of the task (five for each board).*

are predetermined, while placement targets are provided online with respect to the tracked TV. The targets are denoted in Figure 3.8 with dashed lines. A closer look at the placement accuracy is presented in Figure 3.9, which illustrates the y coordinate of the DMP while reaching the moving TV to place the board. The robot starts executing this segment at $t = 14$ seconds and at about 17.5 seconds has almost reached the target over the TV placement location. The tracking error stays below 5 mm during the reaching motion, but a much smaller error of only 0.1 mm is present during the board release at 20.5 seconds. In fact, from 17.5 to 20.5 seconds, the robot is tracking the moving placement target with the proposed DMP formulation with this high accuracy which allows correct board placement. A 3D illustration of the produced DMP trajectory is shown in Figure 3.10. In this figure, it can be seen that the scaling, particularly during the placement, is correctly performed around the rotation axis that is normal to the TV plane.

To evaluate the execution of the demonstrated task, eight people with technical background on the TV assembly cell participated in the user study. The objective of this study is to measure the overall assembly duration and the success rate of the robot in placing the boards. The task of each participant was to collaborate with the robot and fix the boards on the TV frame by screwing after the robot

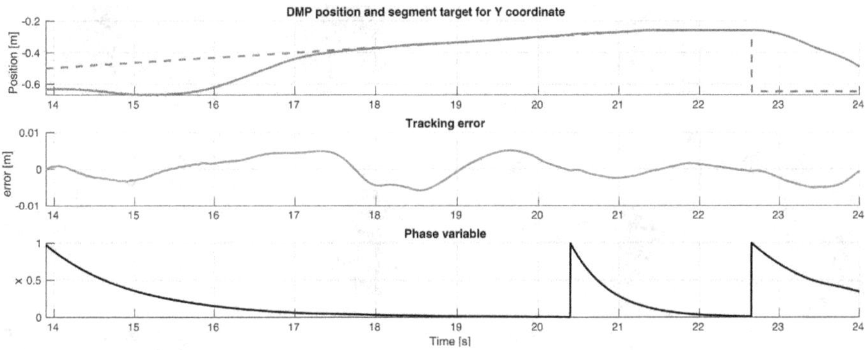

Figure 3.9 *Close-up of Figure 3.8 with the* Y *coordinate of the robot's position during the placement on the moving target (14–22 seconds). The tracking error of the robot is also illustrated, which is almost zero when the DMP reaches the target.*

has placed a board on the TV, as shown in the executed assembly sequence of Figure 3.11. Each participant was given instructions on the assembly procedure followed by a short training session. Each participant performed three assembly cycles. Moving target estimates provided by the perception system could have errors up to 1 cm due to the reflective surface of the TV back plate. In some instances, these errors required minor adjustments performed by the human in order to snap the board into the pegs before screwing. Nevertheless, the board collaborative assembly on the TV was performed successfully. The average time of all participants to complete the assembly in collaboration with the robot was 69 seconds with a standard deviation of 3.7 seconds. The small standard deviation suggests that the human can collaborate with the robot in a consistent way, with respect to the cycle time.

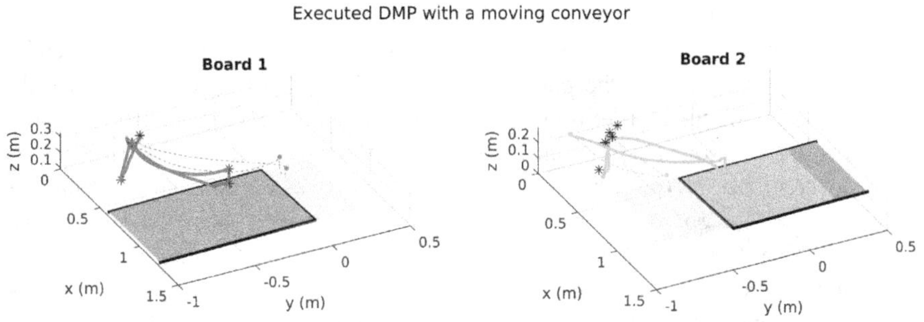

Figure 3.10 *3D figure of the executed DMP with moving conveyor, illustrating the scaling performed because of the moving targets (solid lines) in comparison with the demonstration (dashed lines)*

Figure 3.11 *Collaborative assembly of two electronic boards on a TV frame between a human and the robot*

3.5.3 Evaluation of obstacle avoidance

In this experiment, we demonstrate the effectiveness of the obstacle avoidance method for guaranteeing safety of the human operator. As described in section 1.4.3, for generating the avoidance signal, the manipulator is modeled by virtual capsules. The distance from each capsule is weighted so that it is equal to one for the first two capsules starting from the base and equal to two for the rest in the metric function (3.23). This choice favors avoidance movements of the links that are closer to the human.

Based on the skeleton tracking algorithm, the detected skeletons were modeled by appropriately distributed spheres in order to completely cover the body of the detected humans. The spheres are the dynamic obstacles in the proposed method. When the human worker was close enough to the robot, an additional planar obstacle was added on the TV back plate, to prevent avoidance movements of the robot that would collide with the TV. This obstacle is not required in operation where the human is not in proximity because the scaling presented in section 3.2.2 is such that the resulting trajectory would not pass through the TV.

As shown in Figure 3.12 (top row), the worker moves his hand in the trajectory of the manipulator during the placement of the board and the robot successfully avoids him. While the worker remains in the workspace of the robot, the TV frame planar obstacle remains active, and thus the robot waits for the worker to leave in order to continue. As soon as the worker's hand moves away from the robot, it continues with the board placement. The modified segment is shown in Figure 3.12 (bottom right). In the same figure (bottom left), it is shown that the metric (3.23) stays above zero, which confirms the collision avoidance expected results (no contact with the obstacles). The discontinuities in this figure can be attributed to perception system delays and errors in the detection of the human and the processing to construct and track the skeleton. During the avoidance motion and despite the high robot compliance, the scheme designed in section 3.4.2 ensures successful tracking

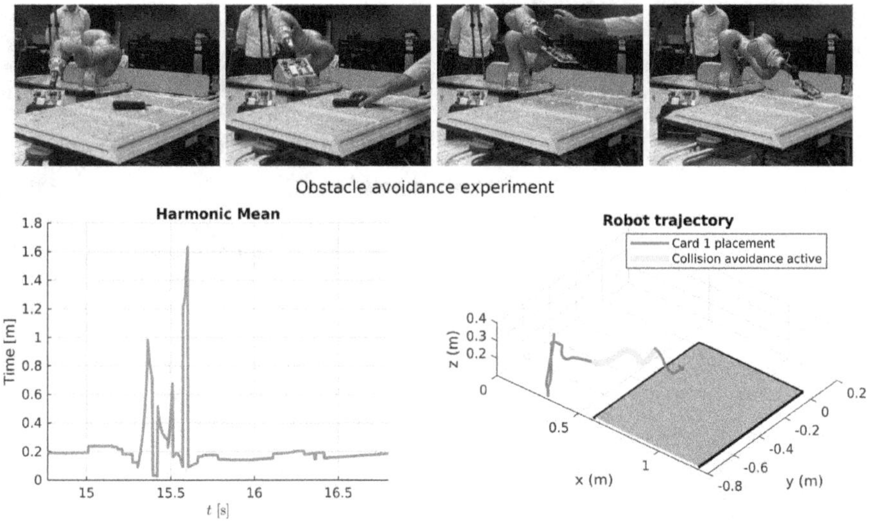

Figure 3.12 *Top: snapshots of the collision avoidance experiment. Bottom: the harmonic mean distance of the robot from the human (left) and the modified trajectory of the robot (right).*

of the modified reference trajectory to avoid the obstacles and eventually execute the desired task.

3.5.4 Evaluation of robot compliance

To demonstrate the second level of safety in case of unexpected collisions with the compliant robot, we deactivated the obstacle avoidance system, simulating a perception system failure. Then, as shown in Figure 3.13 (top), the worker touched the robot with his hand, and when he moved his hand away, the robot successfully continued with the task execution. In the same figure (bottom left) the phase variable is shown to increase until reaching a maximum value based on the increase in the tracking error. This indicates that the execution of the DMP was paused while the human was intervening with the robot. When the human released the robot, the tracking error returned to small values and the time scaling parameter returned to its initial value, resuming task execution. For the phase variable plot, a logarithmic scale is used to better showcase the exponential rise and drop of τ through the sigmoid function (3.13). The modified path is shown in the bottom right of Figure 3.13. Overall, this compliant behavior allowed the human to safely interact with the robot during task execution, limiting the contact forces below 20 N.

3.6 Conclusions

A control scheme was proposed to enable intuitive kinesthetic teaching, adaptive execution, and two-layer human safety, in the form of collision avoidance and compliant

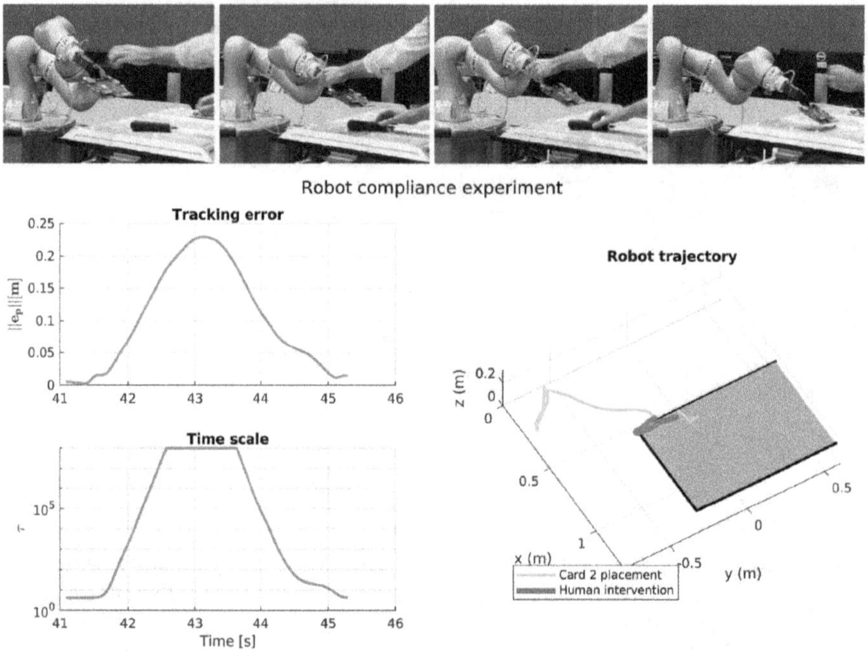

Figure 3.13 *Top: snapshots of the human intervention with the robot being compliant. Bottom: tracking error $\mathbf{e}_p$ norm because of the external force that causes the time scale τ to increase exponentially and pause execution of the phase variable. When the human withdraws his hand, the robot resumes placing the board on the TV.*

behavior. The proposed strategy was successfully integrated into a real-world industrial scenario of a TV assembly. The proposed teaching by demonstration was found to be intuitive by the users who were able to seamlessly train the robotic system to execute the desired assembly task. The robot was able to adaptively execute the learned trajectory to a dynamically changing environment by applying appropriate frame-independent scaling toward a moving target. In all experiments, the target was successfully reached within the predefined accuracy and speed specifications. Safety was realized with the successful integration of obstacle avoidance techniques with compliant control. The manipulator was able to avoid moving obstacles while remaining compliant in case of unexpected collisions owed to perception failure. Both safety layers were thoroughly tested in a variety of experimental scenarios.

Acknowledgments

The research leading to these results has received funding from the EU Horizon 2020 Research and Innovation Programme under grant agreement No 820767, project

CoLLaboratE. The authors would like to thank Turgut Koksal Yalcin and the rest of Arcelik Atolye 4.0 team for hosting the experimental evaluation in their facilities.

References

[1] Ajoudani A., Zanchettin A.M., Ivaldi S., Albu-Schäffer A., Kosuge K., Khatib O. 'Progress and prospects of the human-robot collaboration'. *Autonomous Robots*. 2017, pp. 1–17. Available from https://hal.archives-ouvertes.fr/hal-01643655

[2] Bragança S., Costa E., Castellucci I., Arezes P.M. '*A brief overview of the use of collaborative robotsin industry 4.0: human role and safety*' in Arezes P.M., Baptista J.S., Barroso M.P., *et al* (eds.). *Occupational and environmental safety and health*. Cham: Springer International Publishing; 2019. pp. 641–50.

[3] Kumar S., Savur C., Sahin F. 'Survey of human-robot collaboration in industrial settings: awareness, intelligence, and compliance'. *IEEE Transactions on Systems, Man, and Cybernetics*. 2021, vol. 51(1), pp. 280–97.

[4] Billard A., Calinon S., Rdiger D. 'Learning from humans' in *Springer handbook of robotics*. Springer International Publishing; 2016.

[5] Gribovskaya E., Billard A. 'Learning nonlinear multi-variate motion dynamics for real-time position and orientation control of robotic manipulators'. *9th IEEE-RAS International Conference on Humanoid Robots*; 2009. pp. 472–77.

[6] Paraschos A., Daniel C., Peters J., Neumann G. 'Probabilistic movement primitives' in Advances in Neural Information Processing Systems; 2013. pp. 2616–24.

[7] Huang Y., Rozo L., Silvério J., Caldwell D.G. 'Kernelized movement primitives'. *The International Journal of Robotics Research*. 2019, vol. 38(7), pp. 833–52. Available from https://doi.org/10.1177/0278364919846363

[8] Ijspeert A.J., Nakanishi J., Schaal S. 'Movement imitation with nonlinear dynamical systems in humanoid robots'. *2002 IEEE International Conference on Robotics and Automation*; Washington, DC, USA, 2002. pp. 1398–403.

[9] Schaal S., Mohajerian P., Ijspeert A. 'Dynamics systems vs. optimal control — a unifying view' in Cisek P., Drew T., Kalaska J.F. (eds.). *Computational neuroscience: theoretical insights into brain function*. vol. 165. of Progress in Brain Research. Elsevier; 2007. pp. 425–45.

[10] Ijspeert A.J., Nakanishi J., Hoffmann H., Pastor P., Schaal S. 'Dynamical movement primitives: learning attractor models for motor behaviors'. *Neural Computation*. 2013, vol. 25(2), pp. 328–73.

[11] Kastritsi T., Dimeas F., Doulgeri Z. 'Progressive automation with DMP synchronization and variable stiffness control'. *IEEE Robotics and Automation Letters*. 2018, vol. 3(4), pp. 3789–96.

[12] Weitschat R., Aschemann H. 'Safe and efficient human-robot collaboration part II: optimal generalized human-in-the-loop real-time motion generation'. *IEEE Robotics and Automation Letters*. 2018, vol. 3(4), pp. 3781–88.

[13] Dahlin A., Karayiannidis Y. 'Adaptive trajectory generation under velocity constraints using dynamical movement primitives'. *IEEE Control Systems Letters*. 2020, vol. 4(2), pp. 438–43.

[14] Gašpar T., Nemec B., Morimoto J., Ude A. 'Skill learning and action recognition by arc-length dynamic movement primitives'. *Robotics and Autonomous Systems*. 2018, vol. 100, pp. 225–35. Available from https://www.sciencedirect.com/science/article/pii/S0921889017302695

[15] Colomé A., Torras C. 'Dimensionality reduction for dynamic movement primitives and application to bimanual manipulation of clothes'. *IEEE Transactions on Robotics*. 2018, vol. 34(3), pp. 602–15.

[16] Chiaverini S., Siciliano B. 'The unit quaternion: A useful tool for inverse kinematics of robot manipulators'. *Systems Analysis Modelling Simulation*. 1999, vol. 35, pp. 45–60.

[17] Pastor P., Righetti L., Kalakrishnan M., Schaal S. 'Online movement adaptation based on previous sensor experiences'. *2011 IEEE/RSJ International Conference on Intelligent Robots and Systems (IROS 2011)*; San Francisco, CA, 2011. pp. 365–71.

[18] Ude A., Nemec B., Petric T., Morimoto J. 'Orientation in cartesian space dynamic movement primitives'. *In: 2014 IEEE International Conference on Robotics and Automation (ICRA*); Hong Kong, China, 2014. pp. 2997–3004.

[19] Koutras L., Doulgeri Z. 'A correct formulation for the orientation dynamic movement primitives for robot control in the cartesian space'. *Proceedings of The 3rd Conference on Robot Learning*; Osaka, Japan, 2019.

[20] Koutras L., Doulgeri Z. 'A novel DMP formulation for global and frame independent spatial scaling in the task space'. *In: 2020 29th IEEE International Conference on Robot and Human Interactive Communication (RO-MAN*); Naples, Italy, 2020. pp. 727–32.

[21] Koutras L., Doulgeri Z. 'Dynamic movement primitives for moving goals with temporal scaling adaptation'. *In: 2020 IEEE International Conference on Robotics and Automation (ICRA*); Paris, France, 2020. Available from https://ieeexplore.ieee.org/xpl/mostRecentIssue.jsp?punumber=9187508

[22] Villani V., Pini F., Leali F., Secchi C. 'Survey on human-robot collaboration in industrial settings: safety, intuitive interfaces and applications'. *Mechatronics*. 2018, vol. 55, pp. 248–66.

[23] Arents J., Abolins V., Judvaitis J., Vismanis O., Oraby A., Ozols K. 'Human-robot collaboration trends and safety aspects: a systematic review'. *Journal of Sensor and Actuator Networks*. 2021, vol. 10(3), p. 48.

[24] Stavridis S., Papageorgiou D., Doulgeri Z. 'Dynamical system based robotic motion generation with obstacle avoidance'. *IEEE Robotics and Automation Letters*. 2017, vol. 2(2), pp. 712–18.

[25] Kouris A., Dimeas F., Aspragathos N. 'Contact distinction in human-robot cooperation with admittance control'. *In: 2016 IEEE International Conference on Systems, Man, and Cybernetics (SMC*); Budapest, Hungary, IEEE, 2016. pp. 001951–56.

[26] Deniša M., Petrič T., Gams A., Ude A. 'A review of compliant movement primitives'. *Robot Control*. 2016, pp. 1–17.
[27] Kanakis G.S., Rovithakis G.A. 'Improving safety in human-robot collaboration via dynamic active constraints enforcement'. *2021 30th IEEE International Conference on Robot & Human Interactive Communication (RO-MAN)*; Vancouver, BC, Canada, 2021. pp. 515–19.
[28] Murtaza M.A., Aguilera S., Waqas M., Hutchinson S. 'Safety compliant control for robotic manipulator with task and input constraints'. *IEEE Robotics and Automation Letters*. 2022, vol. 7(4), pp. 10659–64.
[29] Weitschat R., Aschemann H. 'Safe and efficient human-robot collaboration Part II: optimal generalized human-in-the-loop real-time motion generation'. *IEEE Robotics and Automation Letters*. 2018, vol. 3(4), pp. 3781–88.
[30] Hoffmann H., Pastor P., Park D.-H., Schaal S. 'Biologically-inspired dynamical systems for movement generation: automatic real-time goal adaptation and obstacle avoidance'. *In: 2009 IEEE International Conference on Robotics and Automation (ICRA)*; Kobe, 2009. pp. 2587–92.
[31] Park D.H., Hoffmann H., Pastor P., Schaal S. 'Movement reproduction and obstacle avoidance with dynamic movement primitives and potential fields'. *Humanoids 2008 – 8th IEEE-RAS International Conference on Humanoid Robots*; 2008. pp. 91–98.
[32] Ginesi M., Meli D., Roberti A., Sansonetto N., Fiorini P. 'Dynamic movement primitives: volumetric obstacle avoidance using dynamic potential functions'. *Journal of Intelligent & Robotic Systems*. 2021, vol. 101(4), pp. 816–30.
[33] Calanca A., Muradore R., Fiorini P. 'A review of algorithms for compliant control of stiff and fixed-compliance robots'. *IEEE/ASME Transactions on Mechatronics*. 2016, vol. 21(2), pp. 613–24.
[34] Ott C. *Cartesian impedance control of redundant and flexible-joint robots*. 1st ed. Springer Publishing Company, Incorporated; 2008.
[35] Keemink A.Q., van der Kooij H., Stienen A.H. 'Admittance control for physical human-robot interaction'. *The International Journal of Robotics Research*. 2018, vol. 37(11), pp. 1421–44. Available from https://doi.org/10.1177/0278364918768950
[36] Gams A., Nemec B., Ijspeert A.J., Ude A. 'Coupling movement primitives: interaction with the environment and bimanual tasks'. *IEEE Transactions on Robotics*. 2014, vol. 30(4), pp. 816–30.
[37] Vlachos K., Doulgeri Z. 'A control scheme with a novel DMP-robot coupling achieving compliance and tracking accuracy under unknown task dynamics and model uncertainties'. *IEEE Robotics and Automation Letters*. 2020, vol. 5(2), pp. 2310–16.
[38] Koutras L., Doulgeri Z. 'Exponential stability of trajectory tracking control in the orientation space utilizing unit quaternions'. *In: 2021 IEEE/RSJ International Conference on Intelligent Robots and Systems (IROS)*; Prague, Czech Republic, 2021. pp. 8151–58.
[39] Piña E. 'Rotations with Rodrigues' vector'. *European Journal of Physics*. 2011, vol. 32(5), pp. 1171–78.

[40] Bullo F., Murray R.M. 'Tracking for fully actuated mechanical systems: a geometric framework'. *Automatica.* 1999, vol. 35(1), pp. 17–34.
[41] Kastritsi T., Doulgeri Z. 'A passive admittance controller to enforce remote center of motion and tool spatial constraints with application in hands-on surgical procedures'. *Robotics and Autonomous Systems.* 2022, vol. 152, p. 104073.
[42] Stavridis S., Doulgeri Z. 'Bimanual assembly of two parts with relative motion generation and task related optimization'. *In: 2018 IEEE/RSJ International Conference on Intelligent Robots and Systems (IROS)*; Madrid, 2018. pp. 7131–36.
[43] Peng S., Liu Y., Huang Q., Zhou X., Bao H. 'PVNet: pixel-wise voting network for 6dof pose estimation'. *In: 2019 IEEE/CVF Conference on Computer Vision and Pattern Recognition (CVPR)*; Long Beach, CA, USA, 2019. pp. 4561–70.
[44] Peleka G., Mariolis I., Tzovaras D. 'Generating 2.5 D photorealistic synthetic datasets for training machine vision algorithms'. *International Workshop on Soft Computing Models in Industrial and Environmental Applications*; Springer, 2020. pp. 636–46.
[45] Cao Z., Hidalgo Martinez G., Simon T., Wei S., Sheikh Y. 'OpenPose: realtime multi-person 2D pose estimation using part affinity fields'. *IEEE Transactions on Pattern Analysis and Machine Intelligence.* 2019, vol. 43, pp. 172–86.

Chapter 4

Dynamic active constraints construction for enabling human safety during human-robot collaboration

Georgia Peleka[1], Angeliki Topalidou-Kyniazopoulou[1], Ioannis Mariolis[1], and Dimitrios Tzovaras[1]

Collaboration between humans and robots is a critical component of Industry 4.0. Collaborative robots, which combine the incomparable capabilities of humans with the strengths of smart machines to assist operators in performing manual tasks, are currently transforming industrial operations around the world. When working in this environment, occupational health and safety regulations are critical, and active constraints (AC) are a cost-effective and proactive technique to assist robot motion control in achieving human safety while performing collaborative manipulation tasks. A state-of-the-art method for 2D human pose estimation is extended to facilitate accurate human 3D estimation and tracking, which acts as the basis for constructing efficient and resilient AC ensuring safety during human-robot collaboration (HRC).

The AC are described as the surfaces of interest that may cause collisions with the robotic agents participating in the collaborative task. These are the outcomes of tracking human bodies inside the working cell, taking into account context information regarding the collaborative task that is performed by the human worker and robotic agent. The proposed method was deployed in two different setups of industrial tasks that facilitated HRC, and evaluated both qualitatively and quantitatively.

4.1 Introduction

Human employees and robots are the two main workforces of modern industries in the Industry 4.0 spectrum. Human–machine interaction, and more specifically

[1]Information Technologies Institute, Centre for Research & Technology Hellas - CERTH, Thessaloniki, Greece

HRC, is a key cyber-physical and facilitating technology of Industry 4.0 [1]. In a manufacturing context, HRC strives to create an environment in which people may operate alongside robots in close proximity. Humans and robots share the same workspace, resources, and, in certain situations, tasks in such a collaborative setting. The collaboration's primary goal is to combine the robots' power, endurance, repeatability, and precision with humans' intuition, adaptability, and diverse problem-solving and sensory abilities. HRC enables increased overall productivity and improved product quality.

In any HRC system, human safety is a priority. Thus, humans and their robotic collaborators are normally separated, with robots housed in metal cages, which reduces the efficiency and diversity of the manufacturing line. Numerous research attempts on HRC have been made throughout the previous decade. The research describes a variety of ways for supporting multimodal communication, dynamic assembly planning and task assignment, adaptive robot control, and *in situ* operator support. An overall growing movement is evident to lead to the removal of cages so that people and robots can collaborate more effectively. Nonetheless, there are misunderstandings about how robots and people interact: cohabitation, interaction, cooperation, and collaboration. Humans' tasks while interacting with robots are less apparent. Due to a lack of standards and safety measures, the human-robot combination is not widely accepted.

Collaborative robots can be utilized in human-robot teams in diverse manufacturing lines, where robotic manipulators and human employees operate together in a collaborative environment. Human operators can provide superior problem-solving abilities to an HRC team, while robots can supply superior strength and precision. The production system's efficiency may be increased further by utilizing the advantages of both human workers and robots. As a result, human operators will be able to focus on creative activities while robots assist human operators during the assembly. The most critical criterion for an HRC production system to be applicable is the safety of human operators. Maintaining a collision-free environment under all conditions is a fundamental need of any HRC manufacturing system. In the factories of the future, it is projected that there will be an increasing number of interactions between humans and industrial robots [2]. As a result, one of the most important challenges in such scenarios is ensuring human safety.

Toward this end, since the collaborative work cell is not static, we propose a complete pipeline for constructing AC on the fly by obtaining measurements of the working cell and information from the robot's vision system, therefore introducing context-related information to the enforcement of safety algorithms. The main contributions of this work are summarized as follows:

- To extend two state-of-the-art 2D human pose estimation frameworks (Alpha Pose and Open Pose) to accommodate 3D information
- To integrate the 3D information of the acquired human pose into a real-time 3D human tracker

- To represent the 3D human surfaces in the AC space in a more analytical form than 3D points in order for the constraints to be meaningful for successful enforcement
- To assess the developed method in realistic scenarios of HRC in working cells with various setups

The study at hand continues as follows. Section 4.2 outlines work related to the use of AC and context awareness within HRC. Section 4.3 is addressing the four components of our approach. The initial step consists of a 2D detection of people present in the scene and a brief description of the two state-of-the-art methods we examined for the extension. Our extended approach involves identifying 2D skeletons and then projecting them into 3D space, followed by the correction of any miss-detections. This process is used to refine the resulting skeletons. In order to provide smoother performance, the next step is to apply a Kalman filter-based human tracker, which finally leads to the extraction of discrete surfaces that may lead to a risk of collision. These surfaces are presented by a simple primitive shape description, so the constraints that they represent can then be enforced by shaping the impendence of the robot accordingly. Section 4.4 demonstrates the proposed method in the example use cases of a TV assembly and riveting of aircraft parts by collaborative robots, and outlines our experimental evaluation techniques and findings, whereas section 4.5 concludes with an overview and discussion on future research.

4.2 Related work

During assembly tasks, traditional industrial robots overlook human workers and other moving obstacles. Thus, present industrial robot systems are incapable of providing a collision-free workspace for human operators in HRC manufacturing systems. The research community has become more interested in ensuring human protection in the above types of work cells through the principle of collision avoidance, which involves preventing unsafe encounters through the use of protective methods and systems. Some of these methodologies are safety assurance through motion planning and control, context awareness, and human tracking through sensor-based systems [3–5].

Rajnathsing and Li [6] used neural networks to create a monitoring system that can detect the human operator and assure that the robot maintains a minimum safe distance from its human counterpart at all times. Ragaglia *et al.* [7] suggested a new trajectory generation algorithm that maximizes efficiency while taking safety constraints into consideration as actual restrictions. These constraints are formulated by incorporating the manipulator and a series of arbitrarily shaped convex obstacles, which are then fused with depth sensor measurements to describe the area that the human worker will occupy.

Rengevic *et al.* [8] developed a computer vision system that uses state-of-the-art sensors to ensure human safety in the designated safety zones within an automated assembly laboratory. Bdiwi *et al.* [9] created a safety function strategy algorithm for monitoring the entire work cell at different stages of heavy-load industrial robot interaction. The suggested method divides human-robot interaction into steps, with different types of safety functions assigned at each one. Ibarguren *et al.* [10] suggested a security system for industrial robotic cells based on laser rangefinders that can identify and track obstacles around the robot, providing information that can be used to change the robot's behavior and ensure safety.

Context awareness is a relatively new area of HRC research [11–13]. When human operators and robots work together in the same work cell, an entirely new environmental context is formed. This context may include, among other things, information on the assembly environment in general, assembly tools, assembly components, human operator behaviors, and assembly sequences, to name a few examples. Human operators can intuitively detect the context of HRC manufacturing environments by common sense and observations, but robots require sensors and cognition to do so. HRC environments that are both safer and more efficient are possible if robots can comprehend and be aware of their surroundings in the same way that human operators can.

Following the concept of context awareness, we propose a safety constraints construction framework for HRC systems that take into account the context of HRC assembly combined with human posture recognition, thereby ensuring the system's safety and efficiency (Figure 4.1).

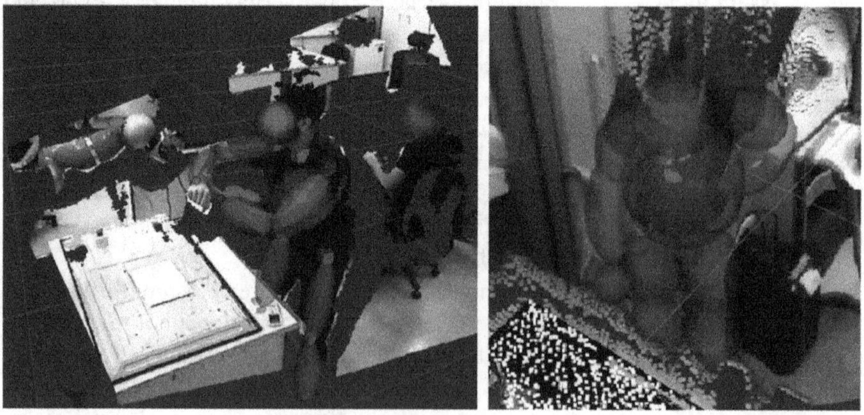

Figure 4.1 Examples of AC depiction through our sensor system in the case where human workers are (left) purposely obstructing the robotic manipulator's path and (right) waiting the robotic manipulator to place a PCB on a TV frame, in order to secure it with screws on the TV frame

4.3 Materials and methods

Our suggested system (Figure 4.2) takes into account context information regarding the collaborative task at hand by utilizing information about occlusions (e.g., during the TV assembly task the human worker's lower body is completely occluded by a conveyor belt), human posture (e.g., the human collaborator will mostly be standing during the task), and the relative positions between the human and the robotic collaborators (e.g., during the TV assembly task, the human and the robot collaborator are positioned on opposing sides of a conveyor belt).

4.3.1 2D Human detection and pose estimation framework

For the initial stage of 2D human detection, we opted to investigate the two state-of-the-art frameworks in this domain: Open Pose [14] and Alpha Pose [15].

4.3.1.1 Open Pose framework

Open Pose is a real-time method for detecting the 2D pose of multiple individuals in a single frame. The bottom-up strategy it employs delivers great accuracy in real time, regardless of the number of persons present in the scene. Open Pose estimates the position of the human skeleton joints as key points on the color image (Red-Green-Blue channels, RGB), for each person that is detected in the scene,

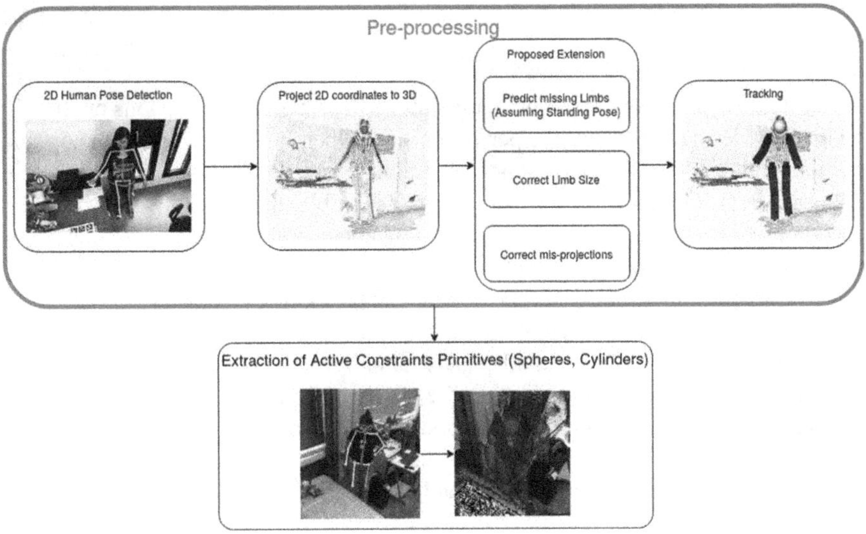

Figure 4.2 The architecture of our proposed system, including the preprocessing steps for 3D human pose extraction (from 2D human detection to the extension to the 3D world, pose correction, and finally to tracking), and the actual constructed AC (using primitive shapes as spheres and cylinders to describe the AC in an analytical form)

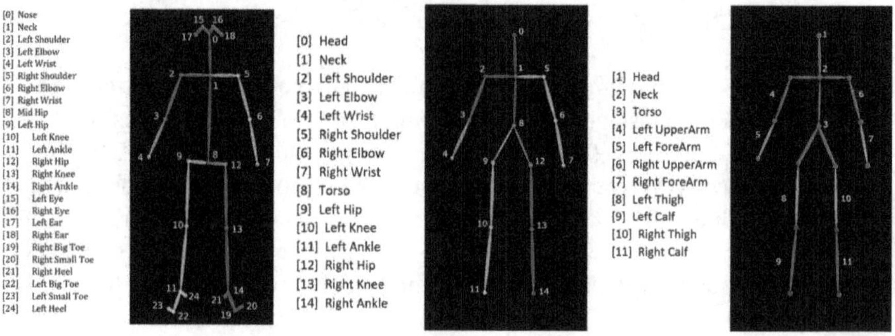

Figure 4.3 *(Left) Mapping of 2D key points in the detected skeleton, using the body_25 model. A total of 25 joints are used to describe each human detected in every image frame [19]. (Middle) Mapping of 2D skeleton key points using the body_25 model's simplification. Each person is defined by a total of 15 joints. (Right) Mapping of the 11 simplified 2D key points to construct the "limbs" representation.*

and provides a confidence value for each of them in real time. The method incorporates three human skeleton models (number and name of joints that describe the human skeleton): MPI[16], COCO[17], and body_25[18]. For this work, from the available human skeleton models, the body_25 model is our model of choice (Figure 4.3, left) since it has the highest detection accuracy.

4.3.1.2 Alpha Pose framework

As we noticed during our early evaluation attempts, when self-occlusions or slightly nonfrontal views are present, the performance of Open Pose degrades. Keeping this in mind and with the objective of attaining a more robust 2D skeleton tracking, the Alpha Pose [15] framework was studied as a feasible alternative for boosting the accuracy and robustness of the initial 2D human detection and, as a consequence, of the subsequent extracted constraints.

Alpha Pose can estimate in real time the 2D poses of several persons in a scene. It adopts a top-down approach, creating a collection of human region recommendations and performing individual pose estimation on each of them; as a result, the execution time scales linearly with the number of humans present in the scene. Because we are concentrating on collaborative human-robot scenarios in which there is essentially only a single human collaborator, we can obtain the quickest feasible inference time by utilizing Alpha Pose.

4.3.1.3 Simplification of the body_25 model

As the first preprocessing step, we are proposing the simplification of the body_25 model in order to streamline the process and increase processing speed. Thus, we are only utilizing 15 joints for the human skeleton (Figure 4.3, middle). After merging their 2D coordinates to approach an average position, the eyes, ears, and

Table 4.1 Comparison of accuracy of Open Pose and Alpha Pose for MPII and COCO datasets

	Mean average precision	
	MPII dataset	**COCO dataset**
Alpha Pose	**82.1**	**72.3**
Open Pose	79.7	61.8

nose are replaced by the "head" joint. Additionally, we opted to disregard any joints below the "ankle" joints and to keep the foot representation as simple as possible by just including the joints of the "right ankle" and "left ankle." The "mid hip" joint is not included in the final calculations, but it is utilized in conjunction with the "neck" joint to form the "torso" joint. The 15 skeleton joints that arise from this process serve as the foundation for the skeleton's "limbs representation" (Figure 4.3, right).

Thus, the "head," "neck," "torso," "right/left upperarm," "right/left forearm," "right/left thigh," and "right/left calf" of each person are also represented as links. This representation of the relation represents the degrees of freedom of the human skeleton that will be used for tracking and, thus, for the actual construction of the AC.

4.3.1.4 Open Pose versus Alpha Pose comparison

After additional testing on the most frequently used pose detection datasets (MPII [16] and COCO [20]), it was discovered that Alpha Pose outperformed Open Pose by a significant margin, achieving an 82.1 mean average precision (mAP) in the MPII dataset and 72.3 mAP in the COCO dataset, where Open Pose reached 79.7 and 61.8 mAP accordingly, as demonstrated in Table 4.1.

A further advantage of Alpha Pose, which is evident in real-world applications, is the method's capacity to cope with occlusions. Figure 4.4 illustrates a visual comparison of Open Pose and Alpha Pose using real-world data with external and self-occlusions. It is clear that Alpha Pose can handle occlusions from the environment better (such as the table on which the TV frame is placed), since it successfully detects the knees and ankles of the person (Figure 4.4, right), whereas Open Pose failed to identify them (Figure 4.4, left). Additionally, Alpha Pose showed superior performance in handling self-occlusions (neck, eyes, and nose joints in Figure 4.4) and with dark-colored areas of the image (right elbow and right wrist joints in Figure 4.4).

4.3.2 Extension for extracting 3D joints

It is critical that skeleton tracking is conducted in 3D, as 2D human detection is inadequate for calculating the possibility of colliding with moving robotic arms. Starting with the detected 2D human skeleton, we propose an extension of 2D

Figure 4.4 Visual comparison of skeletons from Open Pose (left) and Alpha Pose (right) methods in the same image

human pose detection to 3D space, which incorporates two separate steps: projecting the 2D key points into 3D space and a postprocessing step that optimizes the resulting 3D skeleton, considering task context information.

Both of the investigated 2D human pose estimation algorithms (Alpha and Open Pose) can recognize human skeletons and extract their corresponding 3D key points using several RGB cameras. However, the multi-camera method introduces uncertainty into the system due to the necessity for camera-to-camera registration, processing several images concurrently, and camera synchronization. Thus, to obtain the maximum amount of information about the scene, rather than deploying a multi-RGB camera setup for the purpose of extracting 3D key points, we chose to use an RGB-D camera in conjunction with our extension for extracting 3D human skeletons. We extracted a 3D representation of the observed humans by projecting the detected 2D key points to the 3D space using the camera's intrinsic parameters and the registered depth image.

Occlusions, calibration errors, and incorrect detections can all lead to errors and misalignments in the resulting 3D skeletons. Thus, an initial preprocessing stage was employed that analyzes the confidence of each observed joint by assuming that the human worker will be standing during the collaborative task. After being projected to 3D space, any joint with a low confidence level is reassessed and either corrected if possible or "dropped" to the resting position depending on the values of its adjacent key points.

We tested our system with three widely available RGBD sensors: Orbbec's Astra Pro with a range of 0.6 m to 8 m and registered RGB and depth images at 640 × 480 @30 fps (Astra Pro can record up to 1280 × 720 px in RGB, but since we are projecting the 2D points on to 3D, we require registered RGB-Depth pairs), MYNT-EYE P with a range of 0.2 m to 4.2 m, and registered RGB and Depth images at 1280p resolution. These cameras are often used in applications based on the Robot Operating System (ROS), and the specifications above conform to their Ubuntu drivers. The RealSense D435 was chosen for human tracking due to its compact size and weight, which enabled it to be placed closer to the area of

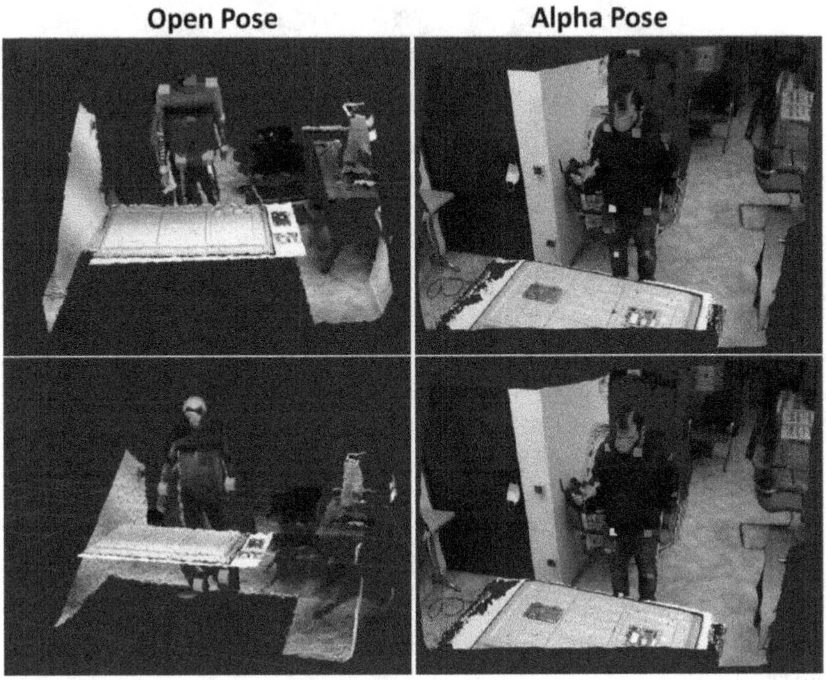

Figure 4.5 *(Top row) The worker's knees and feet are obscured, and hence originally recognized incorrectly on the television frame. (Bottom row) The result of confidence threshold preprocessing. Due to their initial low confidence, the worker's legs are adjusted and they are now in the "standing rest pose."*

operation of the human collaborator. We chose to reduce the resolution of D435 to 640 × 480 pixels, since such resolution is sufficient for identifying and tracking a human in relatively near proximity to the camera, and larger pictures would add additional stress to the system.

We tested our method on the collaborative assembly of a TV, where the worker's legs are typically obstructed by the conveyor belt on which the TV frames are moving. This occlusion usually results in incorrect detection of the person's knees and feet. It is typical for occluded joints to have low confidence and can be recalculated depending on the limb size and the distance between the two shoulders and hips. Figure 4.5 illustrates such instances, in which the worker's legs are incorrectly detected on the conveyor belt and are adjusted by being "dropped" to the floor.

Furthermore, intrinsic calibration errors and wrongful detections can result in incorrect 3D projection with falsely high confidence, as seen in Figure 4.6. To fix such mistakes, an additional preprocessing phase is introduced, which utilizes the skeleton's "limb" representation. The observed human limbs are related to actual limb proportions and adjusted if they are disproportional. The adjustment

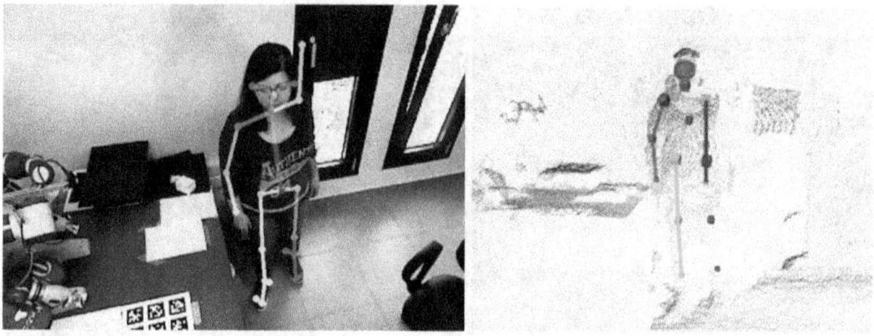

Figure 4.6 *(Left) An illustration of an incorrect detection for the left forearm, due to calibration error, using Open Pose. (Right) Corrected posture of the left forearm, taking into consideration symmetry with the right side and "dropping" the limb to the resting position.*

can address either the scale of the limb in the case of proper detection of the adjacent joints (e.g., for the "forearm," the equivalent joints are "elbow" and "wrist") or the limb can be "dropped" to a resting position under the presumption that the person is "standing" and exploiting the symmetry for the mirrored limb (left/right).

A low confidence value is assigned to all joints that were corrected through the preprocessing steps. As a result, the final 3D skeleton used as an input for human tracking has the original confidence scores for joints that were correctly detected and poor confidence values for all absent or corrected joints. Following this procedure, each skeleton contains all 15 joints of our simplified body_25 model.

4.3.3 3D Human tracking

Starting with the corrected 3D skeleton pose, we employ a standard Kalman filter to predict the movements of human collaborators. We are using our simplified skeleton model of 15 total joints in 3D space, which translates to 3×15 (45) dynamic parameters for 3D position observations and 2×45 (90) dynamic parameters for the state (because we are taking both the position and velocity of each joint into consideration). The model will predict the next state of the human skeleton by using the previously established velocities. The observed 3D coordinates of each skeleton key point are fed into the Kalman filter as input measurements, and their corresponding confidence value is used to calculate the observation noise covariance. The Kalman filter can effectively make predictions since it can give lower noise covariance to observations with high confidence while providing higher noise covariance to observations with low confidence. The next state of the skeleton key points is predicted based on the process distribution and the last observed state, then updated with current observations thus taking the noise error distribution into account.

4.3.4 Active constraints construction

Algorithm 1: Proposed extension algorithm for active constraint construction

```
function ExtractActiveConstraints (rgb_img, depth_img);
Input  : One RGB-D synchronized image pair: rgb and depth images
Output: Vector of active_constraints for every human in the scene
vector< vector< active_constraint >> active_constraints;
vector 2d_skeletons = detectHumansInScene(rgb_img);
vector 3d_skeletons = project2DSkeletonsTo3D(2d_skeletons, depth_img);
for each skeleton ∈3d_skeletons do
    for each joint ∈skeleton do
        check_depth_validity;
        check_limb_size;
        fix_miss_alignments;
    primitive_representation = create_primitive_representation(skeleton);
    vector constraint_vector =
      extract_active_constraints(primitive_representation);
    active_constraints ← constraint_vector;
return constraint_vector
```

The human skeleton must be depicted in a more analytical form than 3D points in order for the constraints to be meaningful for successful enforcement. In order to represent the human surfaces in the AC space, simple primitive shapes such as spheres and cylinders were chosen. In the context of the collaborate project [21], the AC construction module was integrated and evaluated in conjunction with the corresponding module for their enforcement (Figure 4.7).

When looking for the best fitting shapes to describe the human body, we tried several approaches, such as portraying both limbs and joints with spheres or cylinders. Based on the limb's particular geometric form, the most fitting shape is a mixture of spheres and cylinders that are described by their 3D center location, orientation, and scale along all axes. In this manner, the human body can be completely "covered" but still not being over-inflated, resulting in a major decrease in the robots' workspace.

During the evaluation for this work, we concentrated on two manufacturing assembly tasks: an LCD TV assembly task and the collaborative riveting for the construction of airplane components. Throughout the execution of the TV assembly scenario, the TV frame is transported on a conveyor belt, and two printed circuit boards (PCBs) are placed on it by the robotic agent, while the human partner attaches them with screws. Due to the fact that the robotic manipulator and the human employee are on different sides of the conveyor belt, the task requires only the person's upper body. Thus, taking into account this context information since we have prior knowledge of the HRC scenario, AC resembling the upper body are prioritized above those of the lower body.

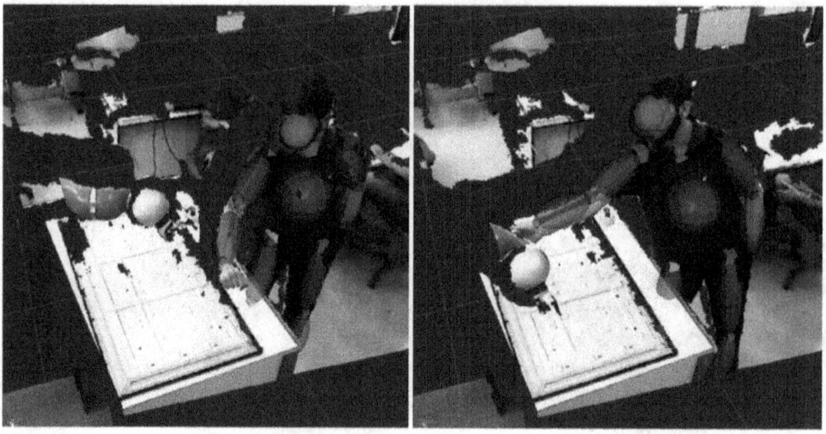

Figure 4.7 AC construction and AC enforcement are being evaluated as an integrated solution. The robotic manipulator effectively avoids the human worker by utilizing primitive shape perception to determine the amount of space occupied by the person in its workspace.

For the assembly of the airplane components, once the assembly is commenced, an automated guided vehicle (AGV) bearing the robotic manipulator navigates to a location where the float may be reached. The worker is positioned directly across the float. After identifying the rivet for riveting, the manipulator positions the bucking bar on the rivet. The worker detects the force change while applying a modest constant force to the rivet using a pneumatic hammer. Once the hammer is activated, the worker feels the force change and applies the opposing force, which completes the riveting process. Because the robotic manipulator and the human employee are on opposite sides of the float, the human body is nearly totally blocked by the floating structure. The lower body of the worker is prioritized above the upper body since it is the one that the AGV can reach.

4.3.5 Fine-tuning of human tracking

A set of fine-tuning steps has been included to further improve the robustness of human tracking and constraint generation. To begin, trials with camera positioning were conducted to assure that the 2D human pose estimation algorithms had the best feasible view. A less frontal but somewhat sideways perspective has been proposed since it provides a sharper view of the human operator while decreasing self-occlusions and obstructions caused by the robotic manipulator.

Furthermore, the Kalman filter parameters have been fine-tuned to adapt to the particular context in order to increase tracking performance. Additionally, there were instances where the robot space was confined by the torso constraint of the human collaborator, such as when the human was leading toward the assembly line. To solve this, an alternate cylindrical depiction of the torso link was suggested that more closely resembled the human form. This is not viable when the Open Pose method is employed to estimate the 2D human pose, as the self-occlusion of the hips makes the torso position assumption untenable. When the Alpha Pose approach is used, which is significantly more efficient at handling self-occlusions, the resultant constraints are more efficient and do not reduce the effective robot operational space.

4.4 Evaluation

To evaluate the AC construction, we employed two different approaches: a quantitative method and a qualitative one. A mock-up of the collaborative assembly of a TV was set up for both of them. An additional setup was used for the qualitative evaluation process, where a mock-up setup for collaborative riveting for aircraft parts assembly was used.

4.4.1 Quantitative evaluation

The initial arrangement consisted of two persons in the scenario, who were showing the assembly of a 49-inch flat-screen television. The television was mounted on a static planar board that was accompanied by two PCBs. One person was in charge of placing the PCBs on the TV frame, while the other was responsible of fastening them to the TV frame with screws and an electrical screwdriver. We repeated the experiment twice, and the ground truth for each skeleton in each frame was obtained using manual annotation in both 2D and 3D space (Figure 4.8). The two sequences that were captured were each made up of 71 and 81 frames, respectively. We kept the number of frames and repetitions low since manually annotating two skeletons with 15 joints each in each frame is time- and resource-consuming task.

The testing sequences were built using the 2D skeletons derived from both the Open and Alpha Pose extensions to create the 3D skeletons for tracking. Our approach was evaluated in comparison to the manually annotated ground truth data as well as a plain 3D projection of the original 2D skeletons, in order to determine

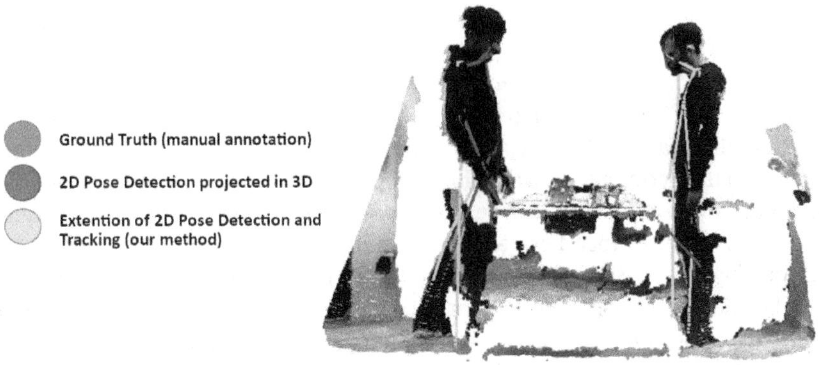

Figure 4.8 The 3D skeletons extracted with our method are being compared with the ground truth (manual annotation) and the simple 3D projection of the initial 2D skeleton

Table 4.2 Quantitative accuracy evaluation results

	Projection in 3D		Extended in 3D		Tracking	
Mean per joint position error	*Open Pose* (cm)	*Alpha Pose* (cm)	*Open Pose* (cm)	*Alpha Pose* (cm)	*Open Pose* (cm)	*Alpha Pose* (cm)
Repetition #1 (71 frames)	13.2	**8.5**	12.9	**8.4**	9.8	**9.6**
Repetition #2 (81 frames)	23.5	**8.3**	27.5	**7.4**	10.5	**8.6**

its overall effectiveness. Our measurement metric was the average 3D Euclidean Distance over all frames and sequences, which allowed us to compute the mean per joint position error per frame and per sequence (in cm).

In scenes like this, where the individual is self-occluded, 2D detections projected to 3D will result in 3D skeletons with missing joints (undetected or with low confidence). Despite the fact that skeletons generated by our proposed extension of the 2D Pose Framework, as well as the tracking approach, had values on all key points (since we estimate all missing joints), only the joints that have a ground truth value were included for this analysis.

When comparing the two 2D Pose estimation frameworks and their performance, we observed several intriguing results throughout the course of this quantitative assessment. Not only does Alpha Pose outperform the Open Pose approach in both the simple 3D projection and extension stages, but it also eliminates the need for tracking, as introducing the tracking layer did not enhance the accuracy of the Alpha Pose extension (Table 4.2).

In general, the extended method delivers the lowest mean per joint position error across both Alpha and Open Pose assessment sequences. Where the basic 3D projection of the original Open Pose 2D skeleton has the highest error in centimeters in both datasets. Accordingly, we may infer from the results that our extension with correction improves the quality of the extracted skeletons, leading to an improved precision in the AC construction process.

However, while the inclusion of a tracking layer using the Kalman filter improved the framework's performance when Open Pose was used for the initial 2D detection, this was not the case when Alpha Pose was employed. As an additional outcome of the performance evaluation of the module's various phases on the inference speed, we were able to derive two alternatives, as shown in Table 4.3. Either the module will use Open Pose for 2D detection and will add extension correction and tracking layers, sacrificing accuracy for speed, or it will employ Alpha Pose with the addition of the extension correction layer, which will result in a compromise between accuracy and speed (sacrificing speed for accuracy).

AC at the current working frequencies of 20 and 15.8 Hz, respectively, are reasonable in either scenario, depending on the operating velocity of the robotic manipulators. Based on a reasonable pace of movement by the human worker within the working cell, these updated rates for the AC are sufficient for maintaining safety.

Table 4.3 *Quantitative speed performance evaluation results*

Projection in 3D		**Extended in 3D**		**Tracking**	
Open Pose	*Alpha Pose* (Hz)	*Open Pose* (Hz)	*Alpha Pose* (Hz)	*Open Pose* (Hz)	*Alpha Pose* (Hz)
20.8 Hz	16	**20**	15.9	**20**	15.8

Also note that, even if we do not use the tracking layer on the top of the Alpha Pose, we can still make use of it to increase the sampling rate of the data. As a result, the Kalman filter may provide estimations in between the Alpha Pose's inferences, allowing for more accurate interpolation of human postures and, thus, faster frame rates. This allows for more consistent management of constraint enforcement.

4.4.2 Qualitative evaluation

The qualitative evaluation was carried out during the CoLLaboratE project [21] integration efforts, where the whole system of AC construction and enforcement was evaluated in two different setups for the TV assembly task, and the AC construction module was evaluated also on the collaborative riveting for the assembly of aircraft parts.

For the software of the AC construction, our hardware setup was:

- A RealSense D435 RGBD sensor
- A workstation with:
  - AMD Ryzen 5 processor
  - GeForce RTX 2080 graphics card
  - 32 GB of RAM

4.4.2.1 Collaborative TV assembly

For the first setup of evaluation and testing, only one human operator and a KUKA LWR 4+ robotic arm were used to replicate an assembly line of HRC. The TV frame was mounted on a flat surface, and the PCBs were positioned on the TV frame by a robotic manipulator while a human worker fastened them with screws. The work cell was shared by both the human and the robotic manipulator, who were working in different directions. The camera was positioned at an angle so that the human partner would not be obscured by the robot as it moved in its workspace. During these experiments, the assembly of the 49-inch TV frame was repeated numerous times for fine-tuning purposes, and it was proved that the robotic manipulator could dependably avoid its human colleague while still successfully completing the task. Figure 4.9 shows an example of avoidance and AC depiction, in which a human worker purposefully obstructs the path of the robotic arm, and it manages to prevent any contact.

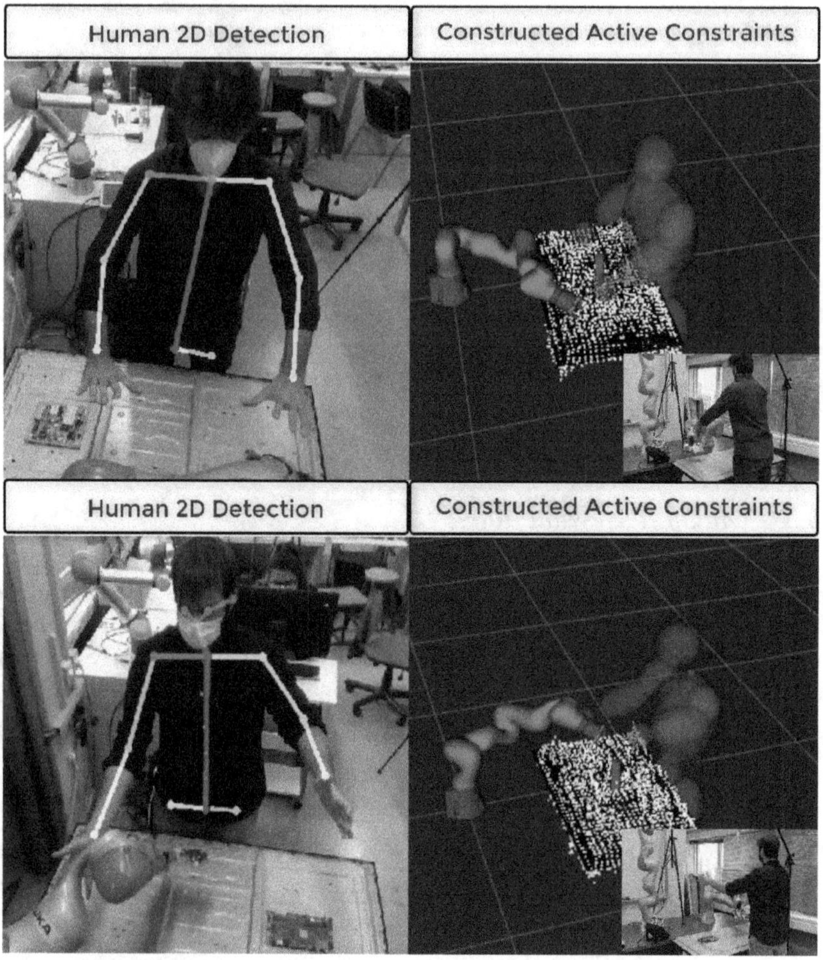

Figure 4.9 *Example of AC depiction in the case of a human worker purposefully obstructing the path of a robotic arm*

Additional integration through in-person meetings was not possible due to coronavirus disease 2019 (COVID-19) restrictions; nevertheless, we conducted numerous qualitative evaluation tests that were similar to the original setup but did not include the robotic manipulator as part of the trial. These additional qualitative trials were crucial in validating the Alpha Pose approach and fine-tuning the overall robustness of the module (Figure 4.10).

4.4.2.2 Collaborative riveting for aircraft parts assembly

For the setup of evaluation and testing on the collaborative riveting task, two human operators were employed to replicate an aircraft parts assembly line. The mock-up

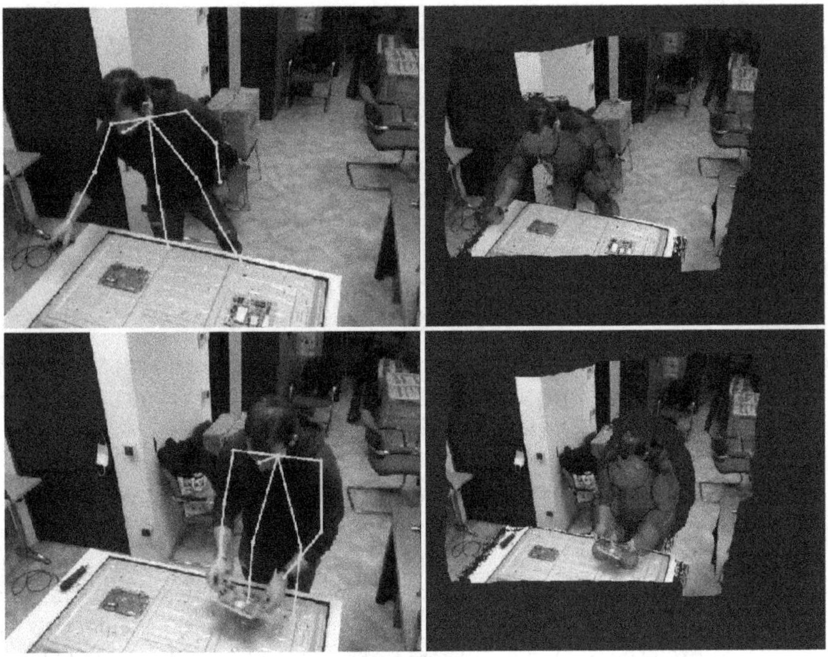

Figure 4.10 *Examples of the experiments performed at the second setup. (Left) The initial 2D detection of the human collaborator using Alpha Pose Framework. (Right) Primitive AC representation.*

float structure frame was mounted on the floor using a metal structure, and the rivets were placed on the float by one operator while the second human worker applied an opposite force with a bucking bar. The camera was positioned at an angle so that the human workers would both be visible and obscured by the float at the minimum degree as they moved in the workspace. During these experiments, the riveting of the airplane float was repeated numerous times for fine-tuning purposes, and it was proved that the human skeletons could be successfully recognized while completing the task. Figure 4.11 shows an example of AC depiction.

4.5 Discussion

On the basis of context awareness, we proposed a method for generating efficient AC for collision-free HRC systems. The primary design objective is to increase the efficiency of the collision-free method based on the proposed extensions of the two state-of-the-art 2D human detection approaches. With the addition of context awareness, the poses of human collaborators can be recognized and in consequence utilized to produce a primitive shape description of the human inside the robotic agents' workspace. Naturally, the robot can be more efficiently controlled to avoid any contact, and the system's performance can be increased.

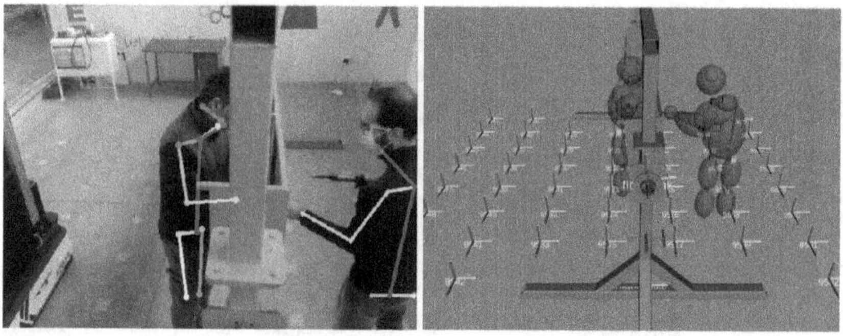

Figure 4.11 *Examples of the experiments performed at the collaborative riveting setup. (Left) The initial 2D detection of the human collaborators. (Right) Primitive AC representation.*

The proposed solution is implemented and integrated with software packages associated with ROS, which for robotics research has been the prevalent choice. Since ROS is open-source, it enables seamless communication between various sensors and robots. On the other hand, this might pose substantial challenges in terms of compatibility, version control, and performance. For example, certain robots can be supported exclusively by particular ROS versions, but some sensors are only accessible in different ones. Additionally, delays and errors can be introduced by the complex communication channels, so the systems reliability and compatibility should be enhanced further.

The operating frequency of vision sensors is a frequent bottleneck for AC construction based on human pose recognition. RGBD sensors now available have a maximum operating frequency of 60 Hz, and in many situations, this is limited to 30 Hz for image sizes that do not put the network under additional strain. The frequency of the RGBD sensor employed in this study was 30 Hz. Our proposed extensions and their associated computational costs resulted in a reduction in the frequency of the constructed constraints to $\sim$20 Hz for the Open Pose framework and $\sim$16 Hz for the Alpha Pose framework.

Even though our achieved working frequencies are reasonable and sufficient for maintaining safety, they heavily rely on the following: (i) a moderate operating speed of the robotic manipulators used in HRC and (ii) a rational pace of movement by the human collaborator within the working cell.

When the operating frequencies are insufficient, the tracking layer may be employed to raise the sample rate of the generated constraints. The Kalman filter may offer approximations in between the inferences made by the 2D human pose frameworks, allowing for more precise interpolation of human positions and hence higher frame rates. This enables more consistent constraint enforcement management.

As an additional point of comparison, even though the Open Pose extension can produce acceptable results with self- and environmental occlusions (especially after the postprocessing steps of correcting the miss-projections and missing joints that we apply prior to the constraint construction), the Alpha Pose outperformed it significantly in the evaluation (see Table 4.2). However, we had to trade quality in order to gain speed, which is why we chose to utilize the Open Pose extension for the final system in the CoLLaboratE project [19]. The goal is to get better inference rates utilizing Alpha Pose in the future after updating our hardware (to a more powerful graphics processing unit - GPU) and optimizing our postprocessing steps. This will allow us to build a system that is more resilient and consistent with occlusions in the future.

4.6 Conclusions and future work

In this work, we proposed extensions to both the Open Pose and Alpha Pose frameworks for human recognition and tracking in 3D space that lead to the construction of AC during HRC. When it comes to preventing collisions between the human and the robotic partners, this strategy has been shown to be quite successful. Currently, the components of the system have been fine-tuned to the extent that a slightly angled side view of the human collaborator can be achieved for all tasks.

The evaluation was performed in the context of CoLLaboratE project, where two test scenarios with three distinct configurations were utilized to illustrate the suggested algorithm's applicability and the overall system's efficacy. The scenarios were based on two use cases of the project: (i) the collaborative TV assembly, where the robotic agent places PCBs on a TV frame, while its human collaborator secures the frame with screws, and (ii) the collaborative riveting in the manufacturing of aircraft components, where the robotic manipulator assists its human counterpart by applying pressure to an area of an airplane float while the human worker performs riveting on it. For the first case, (i) two different setups were used, one involving a human worker and a KUKA LWR4+ robotic arm, and a second complementary one, where the robotic manipulator was omitted from the experiment. This second setup was critical in evaluating the Alpha Pose approach and fine-tuning the overall module's robustness.

As part of our future study, we intend to examine other methods for enhancing the accuracy and overall performance of our proposed system. Additionally, we want to investigate approaches that directly generate the 3D human skeleton, such as VoxelPoseNet [22], and therefore improve performance frequency by eliminating the requirement for preprocessing.

Acknowledgments

This work has been supported by the European Union's Horizon 2020 research and innovation program funded project namely: "Co-production CeLL performing Human-Robot Collaborative AssEmbly (CoLLaboratE)" under the grant agreement with no: 820767. Additionally, we would like to thank the Automation & Robotics Lab of Aristotle University of Thessaloniki for performing the enforcement of the active constraints and providing the lab setup employed in the testing dataset acquisition that included the robotic manipulator.

References

[1] Kagermann H., Wahlster W., Helbig J. 'Securing the future of German manufacturing industry: recommendations for implementing the strategic initiative Industrie 4.0'. *Final Report of the Industrie*. 2013, vol. 4.

[2] International Federation of Robotics, IFR Publishes Collaborative Industrial Robot Definition and Estimates Supply. Available from https://ifr.org/post/international-federation-of-robotics-publishes-collaborative-industrial-rob

[3] Wang L., Givehchi M., Adamson G., Holm M. 'A sensor-driven 3D model-based approach to remote real-time monitoring'. *CIRP Annals*. 2011, vol. 60(1), pp. 493–96.

[4] Morato C., Kaipa K.N., Zhao B., Gupta S.K. 'Toward safe human robot collaboration by using multiple kinects based real-time human tracking'. *Journal of Computing and Information Science in Engineering*. 2014, vol. 14(1).

[5] Zanchettin A.M., Ceriani N.M., Rocco P., Ding H., Matthias B. 'Safety in human-robot collaborative manufacturing environments: metrics and control'. *IEEE Transactions on Automation Science and Engineering*. 2015, vol. 13(2), pp. 882–93.

[6] Rajnathsing H., Li C. 'A neural network based monitoring system for safety in shared work-space human-robot collaboration'. *Industrial Robot*. 2018, vol. 45(4), pp. 481–91.

[7] Ragaglia M., Zanchettin A.M., Rocco P. 'Trajectory generation algorithm for safe human-robot collaboration based on multiple depth sensor measurements'. *Mechatronics*. 2018, vol. 55, pp. 267–81.

[8] Rengevic A., Kumicakova D., Kuric I., Tlach V., Drozdziel P. 'Approaches to the computer vision system proposal on purposes of objects recognition within the human-robot shared workspace collaboration'. *Communications Scientific Letters of the University of Zilina*. 2017, vol. 19(2A), pp. 68–73.

[9] Bdiwi M., Pfeifer M., Sterzing A. 'A new strategy for ensuring human safety during various levels of interaction with industrial robots'. *CIRP Annals*. 2017, vol. 66(1), pp. 453–56.

[10] Ibarguren A., Maurtua I., Pérez M.A., Sierra B. 'Multiple target tracking based on particle filtering for safety in industrial robotic cells'. *Robotics and Autonomous Systems*. 2015, vol. 72, pp. 105–13.

[11] Wang P., Liu H., Wang L., Gao R.X. 'Deep learning-based human motion recognition for predictive context-aware human-robot collaboration'. *CIRP Annals*. 2018, vol. 67(1), pp. 17–20.

[12] Wang L., Gao R., Váncza J., *et al*. 'Symbiotic human-robot collaborative assembly'. *CIRP Annals*. 2019, vol. 68(2), pp. 701–26.

[13] Liu H., Wang L. 'Collision-free human-robot collaboration based on context awareness'. *Robotics and Computer-Integrated Manufacturing*. 2021, vol. 67, 101997.

[14] Cao Z., Hidalgo G., Simon T., Wei S.-E., Sheikh Y. 'OpenPose: RealTime multi-person 2D pose estimation using part affinity fields'. *IEEE Transactions on Pattern Analysis and Machine Intelligence*. 2019, vol. 43(1), pp. 172–86.

[15] Fang H.-S., Xie S., Tai Y.-W., Lu C. 'RMPE: regional multi-person pose estimation'. *2017 IEEE International Conference on Computer Vision (ICCV)*; Venice, 2017. pp. 2334–43.

[16] Andriluka M., Pishchulin L., Gehler P., Schiele B. '2D human pose estimation: new benchmark and state of the art analysis'. *Proceedings of the IEEE Conference on Computer Vision and Pattern Recognition*; Columbus, OH, USA, 2014. pp. 3686–93.

[17] Lin T.-Y., Maire M., Belongie S., *et al*. 'Microsoft COCO: common objects in context'. *European Conference on Computer Vision*; 2014. pp. 740–55.

[18] Zhou C., Li W. 'Pose comparison based on part affinity fields'. *2019 IEEE 2nd International Conference on Information Communication and Signal Processing (ICICSP)*; 2019. pp. 519–23.

[19] *OpenPose github repository*. Available from https://github.com/CMU-Perceptual-Computing-Lab/openpose/blob/master/doc/02_output.md

[20] LinT.Y., Maire M., Belongie S, *et al*. 'Microsoft coco: common objects in context'. *European Conference on Computer Vision*; Springer, 2014. pp. 740–55.

[21] *Co-production CeLL performing human-robot collaborative assembly (collaborate)*. Available from https://cordis.europa.eu/project/id/820767

[22] Zimmermann C., Welschehold T., Dornhege C., Burgard W., Brox T. '3D human pose estimation in RGBD images for robotic task learning'. *2018 IEEE International Conference on Robotics and Automation (ICRA)*; Brisbane, QLD, Australia, IEEE, 2018. pp. 1986–92.

Chapter 5

Time-to-contact for robot safety stop in close collaborative tasks

Alberto Olivares-Alarcos[1], Sergi Foix[1], and Guillem Alenyà[1]

Within the next years, industrial collaborative robots will collaborate with human operators, whose safety might be compromised. To ensure a safe collaboration, robots should estimate the risk of collision with respect to the pose and motion of operators within the shared workspace. A common approach is to compute and maintain a minimum distance between humans and robots during tasks' execution. Nevertheless, separation-based solutions do not capture the real dynamics of the human-robot interaction and tend to be rather conservative, avoiding a close human-robot collaboration. In this chapter, we explore the concept of time-to-contact (TTC) as a softer trigger of safety stops in collaborative scenarios where humans and robots are in constant closeness. Particularly, we propose a TTC formulation and study its advantages with respect to two approaches based on the protective distance proposed by the ISO standards. We compared the three methods in some representative cases extracted from an example of a collaborative task. The evaluation is firstly done in simulation and then in a more realistic setup with a simulated human, aiming for repeatability, and a real robot. Furthermore, we showcased our approach in a demo of a complete collaborative task. TTC allows robots to operate closer to humans and for longer times before a safety stop is issued, which benefits long-term productivity. This later stop produces shorter human-robot distances, which might affect safety. However, the increment in time that the robot moves before stopping (productivity) is far greater than the reduction in the distance (safety). Hence, we can state that TTC greatly improves productivity while slightly compromising safety. In conclusion, our work demonstrates that TTC is a smoother, but still safe, collision risk estimator for close human-robot collaboration.

[1]Institut de Robòtica i Informàtica Industrial, CSIC-UPC, Barcelona, Spain

5.1 Introduction

Over the last few years, there has been a growing need in the industrial sector for more flexible manufacturing processes where humans and robots are expected to work together. Nevertheless, freeing robots from their restrictive current work cells would compromise humans' safety. Collaborative robots, or cobots, are robots specifically designed for direct interaction with humans within a defined collaborative workspace. Collaborative robots shall, among others, be able to perceive human operators' pose and motion, estimate possible risks, and adapt their behavior to ensure humans' safety and reliable and productive collaboration.

In 2011, the International Organization for Standardization released the ISO 10218.1 [1] and the ISO 10218.2 [2], which presented safety guidelines for industrial robots. In 2016, the ISO/TS 15066 [3] extended the previous standards providing specific guidance for safety in collaborative robotics, where a formulation to compute the minimum protective distance was proposed. One of the main limitations of this formulation is that the real direction of motion of the robot and the human is not taken into account. Hence, it results in an over-conservative risk estimator and prevents proper collaboration in applications where humans and robots constantly and closely share the workspace. Indeed, there is no standard way to address this issue yet, and the actual implementation of the formula is still greatly left to the discretion of the integrator [4].

Inspired by the aforementioned ISO standards, several works about safety in collaboration have been published during the last few years [5–8]. Indeed, some of them discussed and aimed to overcome different ISO's formulation drawbacks [9, 10]. However, there is still room for improvement, especially in collaborative tasks in which the human-robot closeness is regular (such as the one in Figure 5.1). Hence, our aim is to explore an alternative to the ISO's formulation well adapted to intensive human-robot collaboration.

In this chapter, we explore the concept of time-to-contact (TTC) [11] as an indicator of the likelihood of the risk of collision during close human-robot collaborations. We show how TTC can be computed and used to stop the robot to avoid collisions while closely sharing the execution of a task with the human. Unlike most of the approaches within the literature, TTC naturally captures the dynamics of the human-robot interaction, considering the actual pose, speed, and direction of both agents. This makes TTC a more precise risk indicator, delaying safety stops and allowing the robot to operate for longer times before stopping. The contributions of this work are:

- an extension of a two-dimensional formulation to compute TTC in a three-dimensional space;
- an algorithm to stop the robot's motion based on the computed TTC;
- a two-fold evaluation, first, in simulation, and second, using a real robot and a simulated human (aiming for repeatability). In both, using prototypical motions to validate the improvement of our approach in delaying the safety stop compared with two ISO-based methods;

Figure 5.1 *Collaboratively filling a tray: example of an industrial task where the human and the robot continuously share both the workspace and the execution of the task (pick and place)*

- a qualitative validation through the implementation of our approach in a real use case: a complete collaborative task in which a human and a robot, closely interacting, filled a tray with tokens.

5.2 Related work

The ISO standards for safety in collaborative robotics [1–3] proposed to use the speed and separation monitoring (SSM) approach, maintaining a robot's speed and a minimum distance between the robot and the human. SSM has been extensively applied in the collaborative robotics domain, and it is utterly aligned with the scope of this work. In some works, the robot's speed is adapted to the distance using discrete regions of space [5–7]. Aiming to enhance the collaboration's fluency, several authors proposed to continuously adapt the robot's speed with respect to (w.r.t.) the current human-robot distance, but still without considering the human's and robot's motion direction [8, 12–14]. Looking for further improvement, more complex formulations to compute the human-robot distance considered the human's and/or the robot's motion direction [9, 10, 15]. In all those articles, as in our approach, the robot's behavior is adapted based on the estimation of a possible risk of collision with a human. The estimation is based on the distance between the human and the robot, and the robot adapted its speed. In our work, we propose to simplify the formulation using the TTC concept, as it naturally embeds the directions and velocities of the two agents.

TTC has been widely used in the literature for automotive collision estimation, warning, and avoidance [16–19]. Authors proposed different approaches: vehicle-to-vehicle and vehicle-to-driver warning, obstacle avoidance, autonomous emergency braking system, etc. In those works, TTC was computed in a two-dimensional space since cars move in a plane. In this work, we explored how the ideas discussed in those articles might be extrapolated to collaborative industrial scenarios, where humans and robots move in a three-dimensional space.

We have opted for stopping the robot because that is the most compliant strategy with the regulations of industrial environments (e.g., ISO). However, TTC has also been used to control other robot's reactions such as adapting the speed or modifying the robot's motion plan for different applications: robot docking and landing [20, 21], Unmanned Aircraft System maneuvers [22], and obstacle avoidance and target chasing [23]. Those works presented interesting approaches showing the potential use of TTC in different applications, which might be considered as ideas for our future work.

5.3 TTC-based safety stop for close human-robot collaboration in industrial tasks

5.3.1 Background

TTC is a biologically inspired measure typically used for obstacle detection and reactive control of motion. It can be defined as the time that an observer will take to make contact with a surface assuming constant relative velocity. Hence, TTC is usually expressed in terms of the speed and the distance of the considered obstacle:

$$ttc = -\frac{Z}{\frac{dZ}{dt}} \tag{5.1}$$

where Z is the distance between the observer and the obstacle, and dZ/dt is the velocity of the observer w.r.t. the obstacle. It is possible to compute TTC from a pure computer vision perspective by just detecting the deformation of objects in consecutive RGB images without calibration [24, 25]. These methods are very sensitive to error detection, so we will compute TTC utilizing the actual observer's and obstacle's poses and velocities as we have access to these measurements. In our work, the observer is the end effector of a robot, and the obstacle is the hand of a human.

5.3.2 TTC computation

Inspired by Hou *et al.* [26] and their TTC formulation for two-dimensional collisions between circles, we propose here a three-dimensional variation of their formula. Hence, the robot's end effector and the human's hand are considered as spheres, and we compute the TTC as the time that it would take the two spheres to collide. Determining when two spheres collide is a matter of determining the moment at which the distance between their centers is equal to the sum of their radii.

Let $\vec{r'} = (r'_x, r'_y, r'_z)$ and $\vec{h'} = (h'_x, h'_y, h'_z)$ denote the positions of the robot and the human, respectively, at the moment of contact. Hence, the distance between them is

$$d = \left\| (r'_x, r'_y, r'_z) - (h'_x, h'_y, h'_z) \right\| \tag{5.2}$$

Knowing that when in contact, the distance is equal to the sum of their radii, and expanding (5.2), we get

$$rad_r + rad_h = \sqrt{(r'_x - h'_x)^2 + (r'_y - h'_y)^2 + (r'_z - h'_z)^2} \tag{5.3}$$

where rad_r and rad_h are the radii of the spheres representing the robot and the human, respectively.

Assuming that the human and the robot are not currently colliding and that both move with constant linear velocity, we can rewrite their positions at the moment of contact based on their current position and velocity:

$$\vec{r'} = \vec{r} + \vec{v}_r ttc \tag{5.4}$$

$$\vec{h'} = \vec{h} + \vec{v}_h ttc \tag{5.5}$$

where $\vec{r} = (r_x, r_y, r_z)$ and $\vec{h} = (h_x, h_y, h_z)$ denote the current positions of the robot and the human, $\vec{v}_r = (v_{rx}, v_{ry}, v_{rz})$ and $\vec{v}_h = (v_{hx}, v_{hy}, v_{hz})$ their current Cartesian velocity, respectively, and *ttc* is the corresponding TTC. Substituting (5.4) and (5.5) in (5.3), we obtain a quadratic equation where, if real positive roots exist, the smallest value is the pursued TTC.

5.3.3 TTC-based safety stop algorithm

Given a TTC between the human and the robot, we can use it so that the robot adapts its current state to avoid possible collisions. In this work, we propose to follow one of the strategies proposed in the ISO 10218.1 [1], issuing a safety-rated monitored stop. Hence, the robot would continue its motion until a certain TTC threshold is violated, from now on, time-to-stop (*ttstop*). In Algorithm 1, we show the decision-making process to compute TTC and stop the robot given that TTC value. If the value of TTC is equal to or smaller than *ttstop*, the robot will stop its motion, and a safety-rated monitored stop will be issued (see line 6 in Algorithm 1). When the value of TTC is greater than *ttstop*, the robot will continue its motion at the task's nominal speed (see line 9 in Algorithm 1). We decided that once a safety stop was issued, the robot would remain stopped and we would exit the TTC-based safety stop loop (see line 2 in Algorithm 1). Since our focus is on very close human-robot applications, the robot would only resume the motion after a human's command. We think that this recovery strategy is the most appropriate one for industrial scenarios similar to our case.

5.4 State-of-the-art approaches: ISO and Fuzzy ISO

In order to evaluate our proposal, we compared it against two methods to trigger a safety stop based on computing the minimum protective distance. First, we used the linear version of the formulation defined in ISO/TS 15066 [3]:

$$S = (v_h T_r + v_h T_s) + (v_r T_r + v_s T_s) + (C + Z_r + Z_s) \tag{5.6}$$

Algorithm 1: decision-making loop to compute TTC, stop the robot and select the robot's speed

Input: Time-to-stop (*ttstop*), nominal robot's speed (V_{r_n}), robot and human radii (rad_r, rad_h)

$safety\_stop \leftarrow false$
while *not(safety_stop)* **do**
  $info_r \leftarrow$ GetRobotPoseVelocity()
  $info_h \leftarrow$ GetHumanPoseVelocity()
  $ttc \leftarrow$ ComputeTTC($info_r$, $info_h$, rad_r, rad_h)
  if $ttc \leq ttstop$ **then**
    $V_r \leftarrow 0$
    $safety\_stop \leftarrow true$
  else
    $V_r \leftarrow V_{r_n}$
  end
  Publish computed robot's desired speed V_r
end

where S is the protective distance, v_h is the "directed speed" of the operator (i.e., the rate of travel of the operator toward the robot), v_r is the directed speed of the robot in the direction of the operator, and v_s is the directed speed of the robot in the course of stopping. T_r is the time for the robot system to respond to the operator's presence, while T_s is the time to bring the robot to a safe, controlled stop. The remaining terms capture measurement uncertainty, where C is an intrusion distance safety margin based on the expected human reach, Z_r is the robot position uncertainty, and Z_s is the operator position (sensor) uncertainty. There is no standard way to measure the human's and the robot's speeds or to compute the times. Indeed, setting the values of the uncertainty constants might also be challenging [4]. Since a discussion of the proper values to choose is out of the scope of this chapter, we will use a simplified version of the formula. First, we will assume that the speed of the robot while stopping, v_s, is equal to the motion speed, v_r. Finally, we will obviate the effect of the uncertainty constants because they would affect all the methods compared in this work. The simplified equation is

$$S = (v_h + v_r)T_b \tag{5.7}$$

where T_b is the sum of T_r and T_s, thus the total time to brake, including the time to respond to the human's presence and to stop the robot. During the different evaluations, we decided that the value used for T_b was also used to set the time-to-stop (*ttstop*). Hence, we established a correlation between our method (see Algorithm 1) and the ones presented in this section. Actually, it makes sense to use the total time to brake as the minimum TTC used to stop (*ttstop*). It would ensure that the robot would stop right before contacting the human, just as the minimum protective distance would.

The second state-of-the-art method used for the evaluations, Fuzzy ISO, was proposed by Campomaggiore *et al.* [10]. They presented a fuzzy-logic system to merge the protective distance formulation with information of the current human's and robot's motion direction. The fuzzy rules are used to scale the effect of the human's and the robot's velocities: for example, when they are going away in opposite directions their method allows them to relax the ISO's formula. The resulting formula would be (5.7) multiplied by the output of the fuzzy-logic system $\alpha \in (0, 1)$:

$$S = \alpha(v_h + v_r)T_b \tag{5.8}$$

Using the aforementioned two equations, we compute the minimum protective distance (S). When the Euclidean three-dimensional distance between the spheres representing the human and the robot is smaller or equal to S, the robot stops, similarly to our proposal (Algorithm 1).

5.5 Validation of TTC for a robot safety stop in close collaborative tasks

We aimed at being able to evaluate all the pros and cons of the different safety stop methods. Hence, we found it necessary to create the taxonomy of all possible situations in tasks such as ours (Figure 5.1). After a comprehensive study, we summarized all possible robot–human collaborative dynamic states in just seven prototypical cases (see Figure 5.2). For the evaluation, we were only interested in the cases that would cause a safety stop. Hence, the cases in which a collision might occur (Figure 5.2(a), (c), and (e)). From now on, we will refer to them as cases A, C, and E, respectively. It is important to remark that we do not need to evaluate the rest of the situations or the complete task execution. The reason is that there will be no change in behavior among methodologies.

5.5.1 Evaluation I – simulation and statistical analysis of the results' significance

First, we evaluated in simulation the performance of our approach against the two methods presented in section 5.4 (ISO and Fuzzy ISO). We used the already mentioned most interesting cases from the taxonomy: A, C, and E.

For each case, we simulated two spheres that moved close to each other. They represented the end effector of a robot and the hand of a human, from now on just robot and human. Once the safety stop was triggered, we finished the simulation. The first hypothesis is that the proposed approach would allow the robot to move for more time before issuing a safety stop. The second hypothesis is that our approach would let the robot get closer to the human but would still be safe, issuing a safety stop before any possible collision.

In order to validate our hypotheses, we evaluated the approaches for several human speeds in each of the three selected situations. Other parameters, such as the human's and robot's initial position or the robot's nominal speed, would affect the triggering of the safety stop. However, we focused on the human's velocity

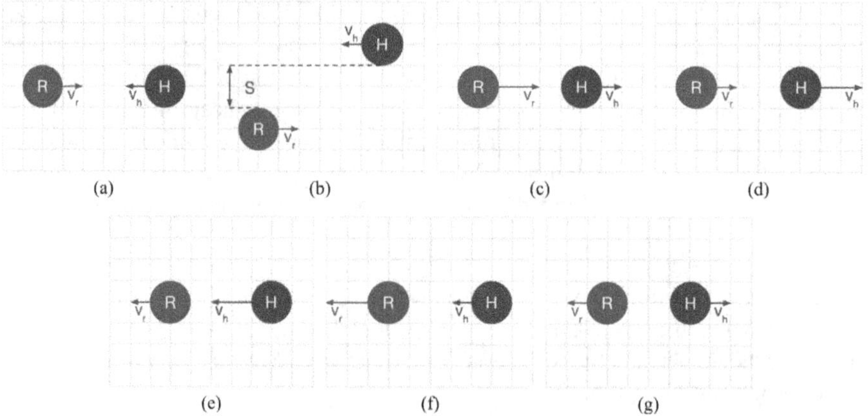

Figure 5.2 *Two-dimensional symbolic representation of the prototypical human-robot collaborative situations. The module of the vector represents the robot's (R) and human's (H) velocity's magnitude (v_r and v_h). (a) Both agents approaching with probable contact. (b) Both agents approaching without probable contact. (c) Robot following the human with probable contact (the robot moves faster than the human). (d) Robot following the human without probable contact. (e) Human following the robot with probable contact (the human moves faster). (f) Human following the robot without probable contact. (g) Both agents getting away. Of all seven cases, only three can trigger a safety stop due to a potential contact: (a), (c), and (e).*

parameter for two reasons. First, it has a direct effect on the dynamics of the interaction, playing a fundamental role in our hypotheses. Second, since it is a parameter we could not control in real scenarios, we wanted to see how the three approaches reacted when it changed. Note that the human and the robot moved on one axis. We randomly generated 1,000 different human's velocities uniformly distributed within an interval of interest for each case. We also added at each simulation timestamp, different white Gaussian noise to the three axes of each of the velocities.

The power of the noise for all three axes of motion was –20 dBW. The noise allowed us to simulate a realistic human motion, not just a straight line movement, as can be seen in Figure 5.3. The simulation time (10 seconds) was selected to match the robot's speed (0.2 m/s) and the maximum reach (1 m) of the robot used in this work, Kinova Gen3. The frequency was 100 Hz, and the *ttstop* (see line 6 in Algorithm 1) and the T_b (see (5.7)) were both set to 0.5 seconds. The radii of the spheres representing the human and the robot were fixed at 0.05 m. The robot's nominal speed was 0.2 m/s. The initial distance between the robot and the human was 0.9 m for case A and 0.4 m for cases C and E. The intervals for the randomly generated human's speeds for the cases A, C, and E were: $[0.3, 0.6]$, $[0.0, 0.15]$, and

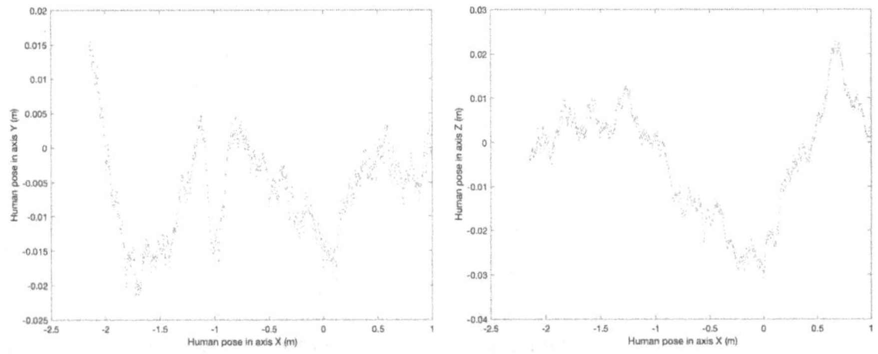

(a) (b)

Figure 5.3 *Example of the human pose evolution for a single simulated noisy motion. (a) and (b) show the evolution of the human pose along the planes XY and XZ, respectively. Note that we simulated a noisy human velocity at each simulation timestamp aiming for a realistic human motion; therefore, the pose evolution did not follow a straight trajectory.*

[0.3, 0.6], respectively. Recall that the motion was mainly in one axis, although we added noise to it. Each case's main direction is depicted in Figure 5.2.

Before a safety stop was issued, we evaluated the time the robot was moving and the final Euclidean distance between the human and the robot. The time before stopping allowed us to validate whether our method delayed the safety stop or not. We reported the distance to show that our method allows the robot to get closer while still being safe (no collision).

5.5.1.1 Evaluation I – first hypothesis: the robot moves more time

We conducted a statistical analysis to evaluate the significance of the proposed approach's improvement in delaying the safety stop w.r.t. the ISO and the Fuzzy ISO methods. We measured the time the robot moved before the safety stop was issued. Figure 5.4 shows the distributions of measured time for each case and evaluated approach.

We manipulated the three different methods (independent variable) and assessed them w.r.t. the time before stopping (dependent variable) for each of the prototypical cases. First, we used Kruskal–Wallis to evaluate if there was a statistically significant difference in group mean, obtaining $\chi^2(2) = 55.08, p < 0.001$, for case A, $\chi^2(2) = 1113, p < 0.001$, case C, and $\chi^2(2) = 1812.8, p < 0.001$, case E. Second, as the results rejected the null hypothesis, we carried out a post-hoc analysis to find out where the differences occurred between the groups. A Dunn and Sidák post-hoc multiple comparison test revealed a significant pairwise difference between our method and the other two in two of the cases: C and E. In case C, the time before braking using TTC was significantly different from the ones using ISO and Fuzzy ISO with a p-value of 0 in both cases. In case E, TTC was also significantly different from ISO

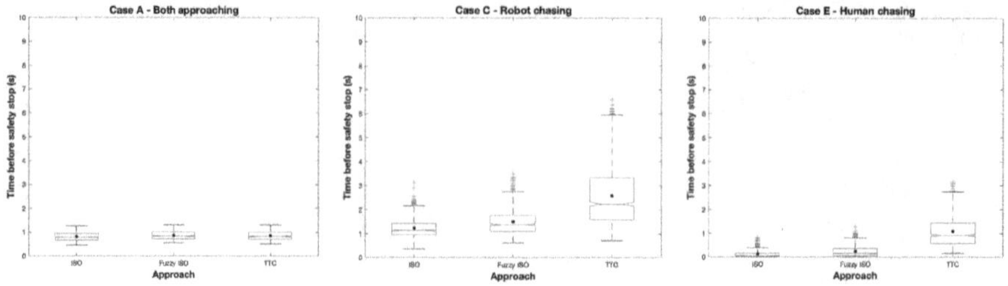

(a) (b) (c)

Figure 5.4 *Distributions of the time the robot moved (a, b, and c) for each simulated case and approach. The mean value is represented by a dot. TTC clearly outperformed the others in (b) and (c).*

and Fuzzy ISO, with a p-value of 0 for both comparisons. These results proved that the differences in the values depicted in Figure 5.4(b) and (c) are statistically significant. Hence, in these two cases, TTC allowed the robot to move for a larger time before stopping. This fact decreased the human-robot protective distance. Specifically, in case C, the robot using TTC moved 110.52% more time on average than with ISO and 71.81% more than with Fuzzy ISO. In case E, the improvement was far greater, 802.68% w.r.t. using ISO and 358.83% w.r.t. Fuzzy ISO. These improvements would be notorious in long-term collaborations at industries, especially in tasks that would imply medium and high levels of interaction (see Figure 5.5). In case A, our method was significantly

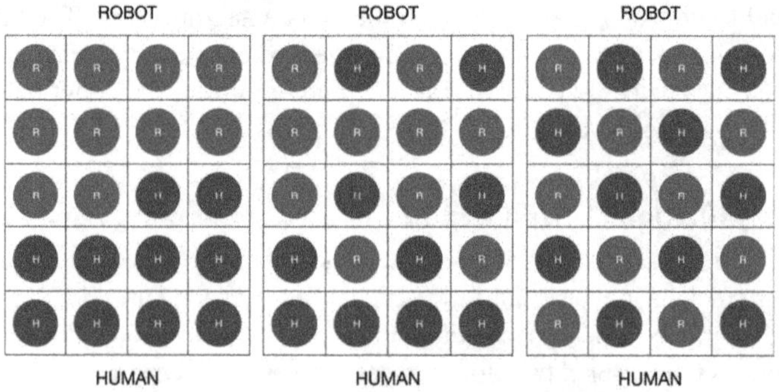

(a) (b) (c)

Figure 5.5 *Examples of our task's distribution between the robot and the human with different levels of interaction. Blue circles (R) indicate the tray's compartments that the robot would fill. Red circles (H) would be filled by the human. (a) Low level of interaction with nearly zero possible crossing trajectories. (b) Medium level of interaction where a few of the robot's and human's targets might involve trajectories' intersections. (c) High level of interaction where the distribution of the task's targets would potentially cause crossing trajectories, leading to several probable contacts and safety stops.*

different from the ISO method with a p-value of $9.010e^{-8}$, while no significant differences were found between the Fuzzy ISO and TTC. However, the mean difference between TTC and ISO was really small (5.85%), as we can see in the values depicted in Figure 5.4(a). Note that in case A, the human and the robot approached each other, which is the default assumption that ISO's formulation does. Hence, we expected our method to capture the high risk of the situation and behaved similarly to the ISO.

5.5.1.2 Evaluation I – second hypothesis: the robot gets closer but it is still safe

A second statistical analysis evaluated the significance of our approach's improvement in allowing closer but still safe human-robot distances. The analysis was again performed w.r.t. the ISO and the Fuzzy ISO methods. We measured the final human-robot distance once the safety stop was issued. Figure 5.6 shows the distributions of measured distance for each case and evaluated approach.

We manipulated the three different methods (independent variable) and assessed them w.r.t. the human-robot distance after stopping (dependent variable). This is for each of the three prototypical cases. First, we used Kruskal–Wallis to evaluate if there was a statistically significant difference in group mean, obtaining $\chi^2(2) = 290.71, p < 0.001$, for case A, $\chi^2(2) = 2138, p < 0.001$, case C, and $\chi^2(2) = 2132.5,\ p < 0.001$, case E. Second, as the results rejected the null hypothesis, we carried out a post-hoc analysis to find out where the differences occurred between the groups. A Dunn and Sidák post-hoc multiple comparison test revealed a significant pairwise difference between our method and the other two in the three prototypical cases: A, C, and E. In case A, the distance after braking using TTC was significantly different from the ones using ISO and Fuzzy ISO with a p-value of 0 and 0.0018, respectively. In cases C and E, TTC was also significantly different from ISO and Fuzzy ISO, with a p-value of 0 for both comparisons.

These results proved that the differences in the values depicted in Figure 5.4 are statistically significant. However, in case A, the mean difference in the human-robot distance produced by TTC was really small w.r.t. the other two methods. Specifically, using TTC the average final distance was 7.69% smaller than with ISO and 2.19%

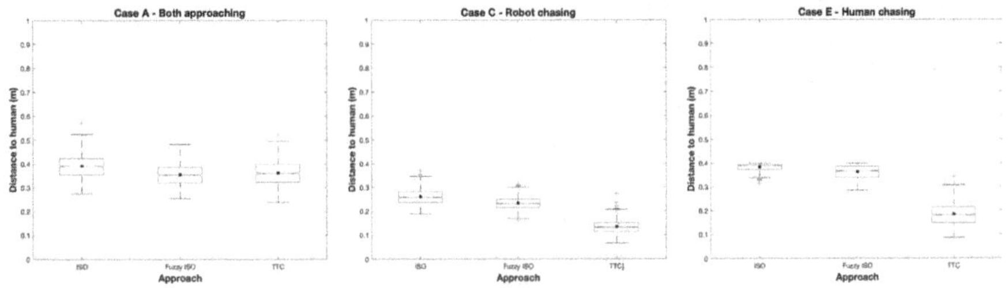

(a) (b) (c)

Figure 5.6 Distributions of the final distance to the human (a, b, and c) for each simulated case and approach. The mean value is represented by a dot. TTC produces shorter distances in (b) and (c) but always ensures safety and avoids collisions.

larger than with Fuzzy ISO. In cases C and E, the differences were larger, and TTC allowed the robot to get closer to the human before stopping. Specifically, in case C, the robot using TTC produced a reduction in the final human-robot of 48.51% and 42.46% w.r.t. ISO and Fuzzy ISO, respectively. In case E, the reduction was a bit larger, 51.82% w.r.t. using ISO and 49.06% w.r.t. Fuzzy ISO. Reducing the final human-robot distance might affect safety, but TTC produced no collisions during the simulation. Furthermore, the percentage reduction in the final distance is far smaller than the increase in the time the robot moves before stopping. Hence, we can say that TTC greatly improves productivity while slightly compromising safety.

5.5.2 *Evaluation II – real robot and simulated human (aiming for repeatability)*

In this case, we contextualized the previous simulation-based evaluation, showing how our proposal may be useful in the task shown in Figure 5.1. The objective was to evaluate our approach implemented in a real robot, avoiding the problem of repeatability of the human operator. Hence, we implemented a realistic setup, where a real Kinova Gen3 robot moved toward a specific target pose. Meanwhile, the system was fed with the position and the velocity of a simulated human. This simulated human moved accordingly to the three cases evaluated before. Specifically, for the situations shown in Figure 5.2(a) and (c), the real robot moved from the grasp pose to the release pose of one of the tray's compartments. This would be the case of a robot picking a token and trying to place it on a compartment. For the remaining situation, Figure 5.2(e), the real robot moved along the opposite trajectory. In this case, emulating when the robot would have already placed the token, and it would go to pick a new one. In this evaluation, we simulated four human's velocities, and we compared the final Euclidean distance from the robot to the target pose after stopping. Note that the distance to the target is related to the two variables studied in section 5.5.1: the robot's motion time and the distance to the human before stopping. Considering the previous evaluation's results, our hypothesis was that the robot would clearly be able to get closer to the target pose using our method in cases C and E. We used the robot Kinova Gen3, equipped with the 2F-85 two-finger gripper from Robotiq, and the same parameters as in section 5.5.1. The four different simulated human's speeds were selected from the intervals we used in section 5.5.1 for the cases A, C, and E: $[0.3, 0.6]$, $[0.0, 0.15]$, and $[0.3, 0.6]$, respectively. Some videos of this evaluation are shown in the additional material.*

Figure 5.7 depicts the evolution of the distance to the target for the four human speeds and each case. These results corroborated, in this case using a real robot, what we already obtained in simulation in section 5.5.1. In case A (see Figure 5.7(a)), the differences between the three approaches remained minimal, around 2%. While in cases C and E (see Figure 5.7(b) and (c)), TTC again outperformed the other two approaches, allowing the robot to get closer to its target pose. Hence, our method lets the robot get closer to finish its task

*www.iri.upc.edu/groups/perception/TTC.

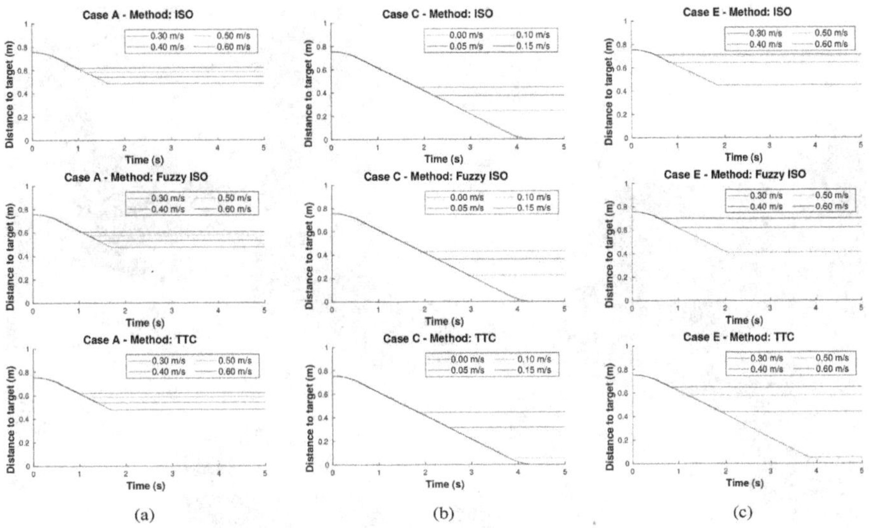

Figure 5.7 *The plots show the evolution of the robot's distance to a target pose during the second evaluation: real robot and simulated human. (a) Both agents approaching with probable contact. (b) Robot following the human with probable contact. (c) Human following the robot with probable contact. Once the safety stop was triggered, the human's simulation finished and the robot remained stopped. This is why the distance to the target becomes constant in the plots, implicitly representing the time the robot was moving before the safety stop. The legend shows the different simulated human's speeds.*

(e.g., placing a token) before stopping. Specifically, in case C, the final distance to the target using TTC was 23.22% and 19.92% shorter on average than with ISO and Fuzzy ISO, respectively. In case E, the improvement is even greater, 31.53% and 29.16% shorter distance to target on average w.r.t. ISO and Fuzzy ISO, respectively. Figure 5.8 depicts a comparison of the final robot's pose w.r.t. the target for the three methods in one of the simulated examples of case E.

5.5.3 Qualitative validation – demo of a collaborative task with a real robot and a human

Finally, we present a qualitative validation of the implementation of our approach in a realistic scenario where a robot and a human share the task of filling the compartments of a tray [27]. The video of one of the task executions can

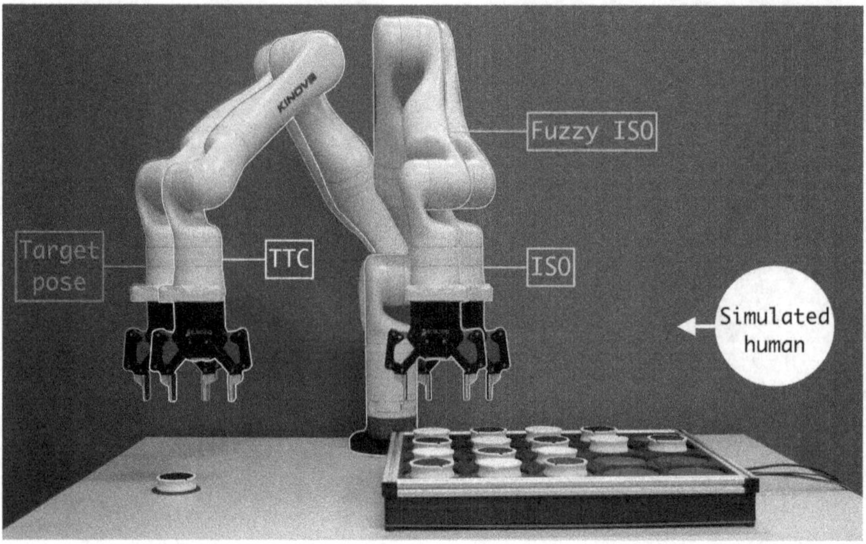

Figure 5.8 *Exemplification of the final distance to the target pose during the second evaluation (case E): real robot and simulated human. The robot is trying to reach the target pose at 0.2 m/s, while the human is following the robot at a faster speed (0.3 m/s in this case). In the image, we can see the target pose (red) and the final robot's pose after the safety stop issued by ISO (cyan), Fuzzy ISO (green), and TTC (yellow). As we can see, TTC allows the robot to get closer to the target before stopping.*

be found in the additional material.* In this implementation, when the human filled one of the compartments the robot was meant to fill, the robot modified its plan and continued with the other free targets. Once a safety stop was issued, the robot was put in joint admittance mode (compliant), and it resumed the motion only after the human's command. The human's pose and velocity were detected using an HTC Vive tracker on their hand and the tokens using an RFID-based (Radio Frequency Identification) board for fast and precise detection. The measurements' rate of the HTC and the robot was 100 Hz. The robot shared with the operator its interpretation of the collision's risk using the lights on the robot's base (see Figure 5.9).

5.6 Conclusion

In this chapter, we studied the concept of TTC to issue a safety stop in close collaborative robotic scenarios. We proposed a TTC formulation and an algorithm to activate a robot safety stop when there is a potential contact. We evaluated our approach against two state-of-the-art methods in a set of prototypical cases. First, in simulation, where a statistical

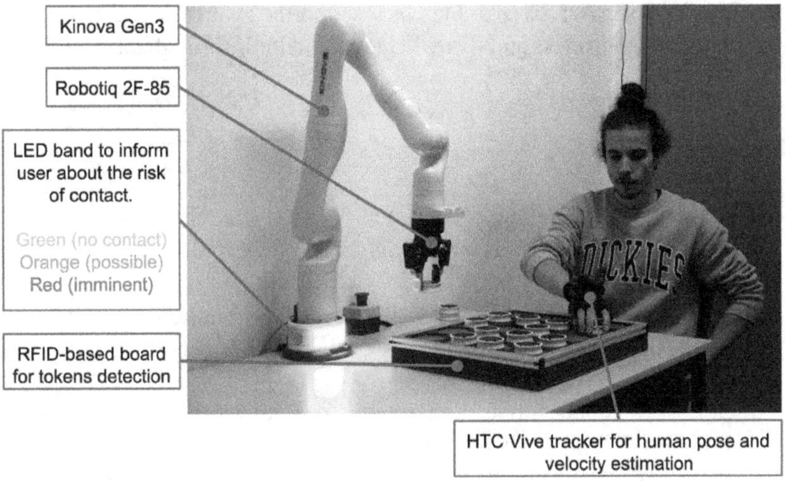

Figure 5.9 Setup for the demo of a collaborative task: filling a tray

analysis was performed to study the significance of the results. Second, with a real robot and a simulated human, aiming for a more realistic evaluation while avoiding the issue of human repeatability. In both evaluations and two out of three cases, our approach clearly produced a later safety stop than the other standard methods, increasing the time the robot is moving/working. In the remaining case, the differences between the three methods were too small to be relevant. Later stops resulted in shorter final distances between the human and the robot. However, the distance was always large enough to avoid collisions. Furthermore, the increment in the time the robot moves before stopping (productivity) is higher than the reduction in the final human-robot distance (safety). This work is a step forward to enabling robots to continue working more time before a safety stop is activated and to get closer to humans while sharing the execution of tasks with them. In the future, we want to extend the quantitative evaluations done in this work to other tasks, robot's speeds, and users. Furthermore, we would like to qualitatively evaluate with users the three compared approaches in terms of perceived safety. We also plan to consider all the human and the robot joints for the computation of TTC. This might be done by reusing and adapting previous work on visual human pose estimation [28]. Finally, it might be interesting to implement an algorithm to continuously adapt the robot's speed using TTC, something that has already shown potential in landing operations for drones.

Acknowledgment

This work has been partially funded by MCIN/ AEI /10.13039/501100011033 under the project CHLOE-GRAPH (PID2020-119244GB-I00); by MCIN/ AEI /10.13039/501100011033 and by the European Union (EU) NextGenerationEU/ PRTR under the project COHERENT (PCI2020-120718-2) and the project ROB-IN

(PLEC2021-007859). A. Olivares-Alarcos is supported by the European Social Fund and the Ministry of Business and Knowledge of Catalonia through the FI 2020 grant.

References

[1] ISO I. '10218 robots and robotic devices-safety requirements for industrial robots – part 1: robots'. *The International Organization for Standardization.* 2011.

[2] ISO I. '10218 robots and robotic devices-safety requirements for industrial robots – part 2: robot systems and integration'. *The International Organization for Standardization.* 2011.

[3] ISO I. 'TS 15066 robots and robotic devices-collaborative robots'. *The International Organization for Standardization.* 2016.

[4] Marvel J.A., Norcross R. 'Implementing speed and separation monitoring in collaborative robot workcells'. *Robotics and Computer-Integrated Manufacturing.* 2017, vol. 44, pp. 144–155.

[5] Vogel C., Elkmann N. 'Novel safety concept for safeguarding and supporting humans in human-robot shared workplaces with high-Payload robots in industrial applications'. *Proceedings of the Companion of the 2017 ACM/IEEE International Conference on Human-Robot Interaction (HRI)*; 2017. pp. 315–16. Available from https://dl.acm.org/doi/proceedings/10.1145/3029798

[6] Nikolakis N., Maratos V., Makris S. 'A Cyber physical system (CPS) approach for safe human-robot collaboration in a shared workplace'. *Robotics and Computer-Integrated Manufacturing.* 2019, vol. 56, pp. 233–43.

[7] Magrini E., Ferraguti F., Ronga A.J., Pini F., De Luca A., Leali F. 'Human-robot coexistence and interaction in open industrial cells'. *Robotics and Computer-Integrated Manufacturing.* 2020, vol. 61, 101846.

[8] Zanchettin A.M., Ceriani N.M., Rocco P., Ding H., Matthias B. 'Safety in human-robot collaborative manufacturing environments: metrics and control'. *IEEE Transactions on Automation Science and Engineering.* 2015, vol. 13(2), pp. 882–893.

[9] Vicentini F., Giussani M., Tosatti L.M. 'Trajectory-dependent safe distances in human-robot interaction'. *2014 IEEE Emerging Technology and Factory Automation (ETFA)*; 2014. pp. 1–4.

[10] Campomaggiore A., Costanzo M., Lettera G., Natale C. 'A fuzzy inference approach to control robot speed in human-robot shared workspaces'. *2019 International Conference on Informatics in Control, Automation and Robotics (ICINCO)*; SciTePress, 2019. pp. 78–87.

[11] Hecht H., Savelsbergh G. '*Tme-to-Contact*' in Vol. 135. Elsevier; 2004.

[12] Lasota P.A., Rossano G.F., Shah J.A. 'Toward safe close-proximity human-robot interaction with standard industrial robots'. *2014 IEEE International Conference on Automation Science and Engineering (CASE)*; IEEE, 2014. pp. 339–444.

[13] Joseph L., Pickard J.K., Padois V., Daney D. 'Online velocity constraint adaptation for safe and efficient human-robot Workspace sharing'. *2020 IEEE/RSJ International Conference on Intelligent Robots and Systems (IROS)*; 2020. pp. 11045–51.

[14] Rosenstrauch M.J., Pannen T.J., Krüger J. 'Human robot collaboration-using kinect v2 for ISO/TS 15066 speed and separation monitoring'. *2018 CIRP Conference on Assembly Technologies and Systems (CATS)*; 2018. pp. 183–86.

[15] Byner C., Matthias B., Ding H. 'Dynamic speed and separation monitoring for collaborative robot applications – concepts and performance'. *Robotics and Computer-Integrated Manufacturing*. 2019, vol. 58, pp. 239–52.

[16] Li Y., Zhang L., Song Y. 'A vehicular collision warning algorithm based on the time-to-collision estimation under connected environment'. *2016 International Conference on Control, Automation, Robotics and Vision (ICARCV)*; 2016. pp. 1–4.

[17] Qu C., Qi W.Y., Wu P. 'A high precision and efficient time-to-collision algorithm for collision warning based v2x applications'. *2018 International Conference on Robotics and Automation Sciences (ICRAS)*; 2018. pp. 1–5.

[18] Song W., Yang Y., Fu M., Qiu F., Wang M. 'Real-Time obstacles detection and status classification for collision warning in a vehicle active safety system'. *IEEE Transactions on Intelligent Transportation Systems*. 2018, vol. 19(3), pp. 758–73.

[19] Yang W., Zhang X., Lei Q., Cheng X. 'Research on longitudinal active collision avoidance of autonomous emergency braking pedestrian system (AEB-P)'. *Sensors (Basel, Switzerland)*. 2019, vol. 19(21), 4671.

[20] Kendoul F. 'Four-Dimensional guidance and control of movement using time-to-contact: application to automated docking and landing of unmanned rotorcraft systems'. *The International Journal of Robotics Research*. 2014, vol. 33(2), pp. 237–67.

[21] Zhang H., Cheng B., Zhao J. 'Extended tau theory for robot motion control'. *2017 IEEE International Conference on Robotics and Automation (ICRA)*; 2017. pp. 5321–26.

[22] Eguíluz A.G., Rodríguez-Gómez J.P., Martínez-de Dios J.R., Ollero A. 'Asynchronous event-based line tracking for time-to-contact maneuvers in UAS'. *2020 IEEE/RSJ International Conference on Intelligent Robots and Systems (IROS)*; 2020. pp. 5978–85.

[23] Kaneta Y., Hagisaka Y., Ito K. 'Determination of time to contact and application to timing control of mobile robot'. *2010 IEEE International Conference on Robotics and Biomimetics*; 2010. pp. 161–66.

[24] Alenyà G., Nègre A., Crowley J.L. 'A comparison of three methods for measure of time to contact'. *2009 IEEE/RSJ International Conference on Intelligent Robots and Systems (IROS)*; 2009. pp. 4565–70.

[25] Garcia A.J.S., Figueroa H.V.R., Hernandez A.M., *et al.* 'Estimation of time-to-contact from Tau-margin and statistical analysis of behavior'. *2016 IEEE International Conference on Systems, Signals and Image Processing (IWSSIP)*; 2016. pp. 1–6.

[26] Hou J., List G.F., Guo X. 'New Algorithms for computing the time-to-collision in freeway traffic simulation models'. *Computational Intelligence and Neuroscience*. 2014, 761047.
[27] Olivares-Alarcos A., Foix S., Borgo S., Alenyà G. 'OCRA – an Ontology for collaborative robotics and adaptation'. *Computers in Industry*. 2022, vol. 138, 103627.
[28] Gassó Loncan Vallecillo J.C., Olivares-Alarcos A., Alenyà G. 'Visual feedback for humans about robots perception in collaborative environments'. *Institut de Robòtica i Informàtica Industrial, CSIC-UPC*. 2020, IRI-TR-20-03.

Chapter 6

A human-centered dynamic task scheduling and safe task execution approach for human-robot collaboration scenarios

Andrea Pupa[1], Mohammad Arrfou[2], Gildo Andreoni[2], and Cristian Secchi[1]

The new paradigm of human-robot collaboration has led to the creation of a shared workspace in which humans and robots collaborate to accomplish a common job composed by a set of tasks. Therefore, the safety standards have been updated, addressing these new scenarios. In this context, a proper integration between a task scheduling strategy and a task execution strategy is crucial for an efficient and natural human-robot collaboration. The first must take into account in real time the variability of the two agents during the collaboration. While the second one must explicitly deal with the safety standards, allowing the robot to execute the tasks in a safe and efficient way.

In this chapter, we propose an integrated framework for task scheduling and execution for human-robot collaboration scenarios. The tasks are dynamically scheduled, dealing with the uncertainty of both agents at runtime. Subsequently, at the task execution level, the robot accomplishes the tasks with a behavior that is forced to be compliant with the safety standards, ensuring the safety of the human operator. Furthermore, the information at this level is mutually integrated with the task scheduling strategy, improving the collaboration. The proposed architecture is experimentally validated in a custom collaborative environment.

6.1 Introduction

The increasing use of collaborative robotics within industrial settings has brought with it notable changes (see, e.g., Reference 1). In particular, the new industrial environments are characterized by the presence of human operators and robots who work closely together, without the aid of physical barriers as in the case of industrial robotics. However, the absence of these barriers has led to a change in the safety paradigm for

[1]Department of Science and Method of Engineering, University of Modena and Reggio Emilia, Italy
[2]Datalogic S.p.A., Italy

the human operators. The safety standards, indeed, have been updated, addressing also these new collaborative workspaces (see References 2 and 3). In particular, the technical specification ISO/TS 15066 [4] provides detailed information on how to assess risks for a Human-Robot Collaboration (HRC) application. Consequently, the researchers have focused on developing strategies that deal with safety for collaborative scenarios (see, e.g., References 5–7).

One of the great advantages introduced by collaborative robotics consists in the possibility of creating a great synergy between two agents, that is, the human operator and the robot, completely different from each other and with complementary skills. Thanks to this synergy, the human-robot team can perform tasks that neither the human operator nor the robot alone could perform. For this, it is essential to understand what is the best way to allocate tasks between the two agents in order to make the most of this collaboration, obtaining all the advantages of a heterogeneous team (see, e.g., Reference 8).

Many works have been done in the field of multiagent task allocation inside the industrial context (see, e.g., References 9–11). However, these works cannot be applied in HRC scenarios, as they are limited to applications with homogeneous agents, which is a rather different paradigm.

For this reason, the researchers also focused on studying the task assignment in the case of collaborative cells. Many works treat and model the problem as a nonlinear optimization problem, optimally allocating tasks between different agents (see, e.g., References 12–14). However, these solutions have a high computational complexity that makes them unsuitable for widely dynamic situations in which it is necessary to reschedule the tasks. In Reference 15, the subtask allocation problem in an HRC scenario is addressed with a two-level feedforward optimization strategy. Moreover, the framework is enriched with an online feedback procedure that exploits mutual trust information to dynamically reallocate the subtasks. In Reference 16, the authors propose a multicriteria decision-making framework to address the task allocation problem. The presented algorithm is capable of handling unexpected events by rescheduling the remaining tasks. In Reference 17, the problem is addressed with a two-level architecture that is capable of dynamically handling task failures. All these proposed solutions assume that both the human operator and the robot always take a constant time to perform the same task, which rarely happens. Several works that consider the dynamic task rescheduling due to the variable human execution time are already available in the literature (see, e.g., References 18 and 19), but they do not explicitly consider the dynamism and uncertainties introduced by the robot. In the case of a collaborative robotics application, in fact, both the robot and the human operator involve uncertainties and must be treated as agents with variable behavior. On the one hand, the real execution time of the human operator is difficult to predict a priori since it is subject to many personal factors, e.g., tiredness. On the other hand, the robot is forced to plan its trajectory online due to the high dynamism of the environment. Sometimes, it may be necessary to change paths to avoid the human operator who moves during the collaboration. At other times, it may be necessary to decrease the speed or, in the worst case, to stop, to ensure that the behavior of the robot complies with the required safety standards. All these possible deviations from nominal behavior must be taken into account in the online rescheduling procedure in order to increase the productivity and performances of the collaboration.

Moreover, to make the overall HRC as natural and efficient as possible, it is necessary not only to improve the mutual awareness at the task scheduling level but also at the task execution level. This improvement could be achieved by enriching the scheduling procedure with a human tracking strategy that monitors the human operator inside the collaborative workspace. This human tracking strategy can be exploited at the scheduling level to estimate online the real execution times of the tasks and at the task execution level to ensure safety.

In this chapter, we propose a novel dynamic framework for HRC that mutually integrates the task scheduling and the task execution strategies. The proposed architecture is completely aware of the activity of the human inside the collaborative workspace and exploits this information to ensure safety during the robot task execution. Moreover, the safety is mutually integrated also at the scheduling level, improving the dynamic rescheduling procedure. The proposed framework is composed of two layers. The first layer is responsible for generating the schedule for the human and the robot. Considering the nominal task durations, it first computes the optimal nominal schedule. Subsequently, it exploits both the human tracking strategy and the planning information to reschedule online the tasks, adapting the optimal schedule. The second layer, instead, takes care of the execution of the tasks while ensuring safety for the human operator. It first computes the robot trajectory, replanning it online, if necessary, and then it scales the robot velocity to make the robot behavior compliant with the ISO/TS 15066 standard. The chapter extends the previous work proposed in Reference 20, adding a more in-depth discussion to the results obtained.

The main contributions of this work are:

- A dynamic task scheduling strategy that exploits both the real task execution time and the human behavior to dynamically reschedule the tasks, improving the collaboration between the human and the robot.
- A dynamic task execution strategy that is computationally cheap but still manages to generate the optimal input for the robot that complies with safety standards. Additionally, this strategy can provide feedback in the event that it is impossible to perform a task.
- A novel adaptive framework that mutually integrates the two strategies, making the collaboration more natural and efficient.

The paper is organized as follows: In section 6.2, the task scheduling and the task execution problem for a collaborative cell are presented, while in section 6.3, the speed and separation monitoring collaborative mode is treated. Section 6.4 presents the integrated proposed architecture. In section 6.4.1, an algorithm to address the dynamic task scheduling is detailed, while in section 6.4.2, the safety dynamic task execution procedure is presented. In section 6.5, an experimental validation of the proposed architecture is presented, while in section 6.6, a depth discussion is presented. Finally, in section 6.7, some conclusions and future works are addressed.

6.2 Problem statement

We consider a collaborative industrial workspace, where two agents, a robot manipulator R and a human operator H, have to collaborate in order to achieve a common goal, i.e., executing a collaborative job divided into a set of tasks $(T_1, \ldots, T_N)$. Each task is defined by an intrinsic cost w_{ai} and a nominal execution time t_{ai}, where $a \in A = \{H, R\}$ represents the agent that executes the task i. However, during the collaborative work, the two agents may deviate from the planned behavior. The operator, in fact, is in itself a highly unpredictable agent since it is unlikely that it will always take the same amount of time to perform the same task. The robot, on the other hand, has to adapt to this unpredictability, generating online the best trajectories to perform its tasks. These trajectories must be collision-free, e.g., avoiding the human operator, and must comply with safety regulations, which impose an upper bound for the robot velocity, depending on the distance with the human operator. For this reason, the workspace is equipped with a human monitoring unit (see, e.g., References 21–23) that is exploited to estimate online the real execution time of the human operator (see, e.g., References 24–26) and to adapt in real time the robot behavior, planning a trajectory which is always collision-free, thus feasible, and compliant with the safety standards.

Considering a collaborative robot manipulator with n joints, this must accomplish a task that requires moving from an initial configuration $q(t_i) = q_i \in \mathbb{R}^n$ to a desired final configuration $q(t_f) = q_f \in \mathbb{R}^n$. Once defined the best trajectory $q(t)$ that the robot has to perform, it is possible to explicitly isolate the magnitude of the velocity by applying a path-velocity decomposition:

$$q(t) = q(s(t)) \qquad t \in \left[t_i, t_f\right] \tag{6.1}$$

where s is the curvilinear abscissa that parameterizes the geometrical path $q(s)$, while the variation of s represents the time law of the desired path, e.g., the velocity profile along the desired path.

Differentiating (6.1), it is possible to obtain:

$$\dot{q}(t) = q'(s)\dot{s} \qquad t \in \left[t_i, t_f\right] \tag{6.2}$$

where $q'(s)$ is the vector tangent to the desired path, while $\dot{s}$ constitutes the magnitude of the robot velocity.

To be feasible, the trajectory $q(t)$ must avoid the human operator during the collaborative job. Defining m as the number of the human limbs, the trajectory $q(t)$ is considered collision-free when:

$$d(\sigma_{Ri}(q(t)), \sigma_{Hj}(t)) \geq d_{\min} \qquad \begin{array}{l} \forall i \in \{1, \ldots, n\}, \\ \forall t \in \left[t_i, t_f\right], \\ \forall j \in \{1, \ldots, m\} \end{array} \tag{6.3}$$

where $d(\sigma_{Ri}(q(t)), \sigma_{Hj}(t))$ is the distance between the ith robot link and the jth human limb at time t, and $d_{\min}$ is the minimum admissible distance.

In this work, we aim at designing a dynamic and safe task scheduling and execution architecture that:

- Generates the optimal nominal task schedules for the two agents, the human and the robot, i.e., two ordered task schedules that take into account the precedence constraints and the nominal execution times to maximize the parallelism between the human and the robot, enabling the collaboration as much as possible and minimizing the job makespan.
- For each robot task finds a set of trajectories that are always collision-free and compliant with the limits imposed by the ISO/TS 15066 standard, i.e., the computed path avoids the human operator, and the robot velocity is adapted in real time along the path to ensure safety during the collaborative job.
- Dynamically reschedules the tasks considering the real execution times and possible failures in the planning process, i.e., the tasks could be anticipated to speed up the job if the human operator is late or the tasks can be changed if the human operator hinders the robot. This rescheduling procedure improves the performance of the human-robot team by enabling the collaboration as much as possible, always taking safety explicitly into account.

6.3 Speed and separation monitoring

As a consequence of the development of HRC technologies, the safety standards have been updated in order to address the new coworking scenarios. In particular, the ISO 10218-1 and the ISO 10218-2 [2, 3] standards classify the collaborative modes into four different categories: *safety-rated monitored stop, hand guiding, speed and separation monitoring* (SSM), and *power and force limiting*. Moreover, the technical specification ISO/TS 15066 standard [4] dispenses guidelines on how to assess the risk for each of the four collaborative modes.

In modern industrial applications, where a human operator and a collaborative robot share the workspace, the SSM collaboration mode is widely used. In this collaborative mode, the speed of the robot must be adapted according to the position and the velocity of the human operator inside the collaborative workspace. Since the human velocity is not monitored, usually the working area is divided into three different zones considering only the distance between the human and the robot. This scenario is represented in Figure 6.1. When the human operator occupies the green area, the robot is free to move at maximum speed; when the human operator is inside the yellow zone, the robot must operate at reduced speed; and when the human operator occupies the red zone, the robot must stop. The ISO/TS 15066 standard provides the guidelines for calculating the sizes of these areas, namely the minimum protective separation distance S_p, considering also the relative speed between the robot and the human operator. S_p can be computed as

$$S_p(t_0) = S_h + S_r + S_s + C + Z_d + Z_r \tag{6.4}$$

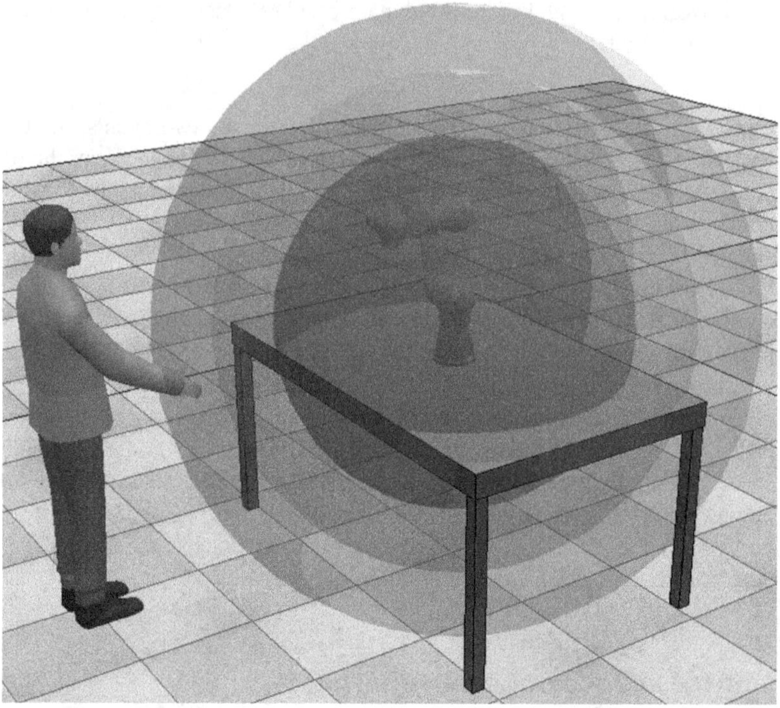

Figure 6.1 *Representation of different safety zones with SSM collaboration mode*

where S_p represents the protective separation distance at time t_0, while t_0 is the current time. S_h, S_r, and S_s represent the contributions to the protective separation distance due to the operator's movement, the robot reaction time, and the robot stopping time, respectively. C, namely the intrusion distance, is the distance that a part of the body can intrude into the sensing field before it is detected. Z_d and Z_r are the position uncertainties of the human operator inside the workspace and of the robot system, respectively.

It is possible to expand the first three terms of (6.4) as

$$S_h = \int_{t_0}^{t_0+T_s+T_r} v_h(t)\,dt \tag{6.5}$$

$$S_r = \int_{t_0}^{t_0+T_r} v_r(t)\,dt \tag{6.6}$$

$$S_s = \int_{t_0+T_r}^{t_0+T_s+T_r} v_s(t)\,dt \tag{6.7}$$

where T_s represents the robot stopping time, and T_r is the robot reaction time. v_h and v_r are the directed speed of the human operator toward the robot and one of the robots toward the human operator, respectively. v_s is the speed of the robot during the stopping phase.

Under the assumptions that the velocity of the robot is constant during the robot reaction time, the acceleration remains constant during the stopping phase, and the dynamics of the human operator is slower than the robot dynamics, which is true in the case of a generic HRC application, (6.5)–(6.7) can be approximated as follows:

$$S_h = v_h(t_0)(T_s + T_r) \tag{6.8}$$

$$S_r = v_r(t_0)T_r \tag{6.9}$$

$$S_s = v_r(t_0)T_s + a_{\max}\frac{T_s^2}{2} \tag{6.10}$$

where $a_{\max}$ is the maximum robot deceleration expressed in as absolute value.

Remembering that the robot stopping time can be expressed as a function of the actual robot velocity:

$$T_s = \frac{v_r(t_0)}{a_{\max}} \tag{6.11}$$

It is possible to further expand (6.8)–(6.10) as

$$S_h = v_h(t_0)\left(\frac{v_r(t_0)}{a_{\max}} + T_r\right) \tag{6.12}$$

$$S_r = v_r(t_0)T_r \tag{6.13}$$

$$S_s = \frac{v_r(t_0)^2}{2a_{\max}} \tag{6.14}$$

Substituting (6.12)–(6.14) in (6.4), it is possible to obtain an upper bound robot velocity:

$$v_{r\max}(t_0) = \sqrt{v_h(t_0)^2 + (a_{\max}T_r)^2 - 2a_{\max}(C + Z_d + Z_r - S_p(t_0))} - a_{\max}T_r - v_h(t_0) \tag{6.15}$$

Equation (6.15) represents the maximum speed that the robot could reach in the direction of the human operator, i.e., the safety constraint imposed by the ISO/TS 15066 standard. Figure 6.2 shows the trend of the velocity limit $v_{r\max}(t_0)$ as a function of the actual separation distance between the human operator and the robot $S_p(t_0)$.

6.4 Architecture

The proposed dynamic and safe task scheduling and execution framework are represented in Figure 6.3. The presented architecture is divided into two different main layers:

1. **The scheduling layer**. It is divided into the **task assignment** block and the **dynamic scheduling** block. The first one creates the nominal optimal schedules for the human operator and the robot based on the maximum parallelism criterion, thus reducing the waiting time. The second one takes into account the

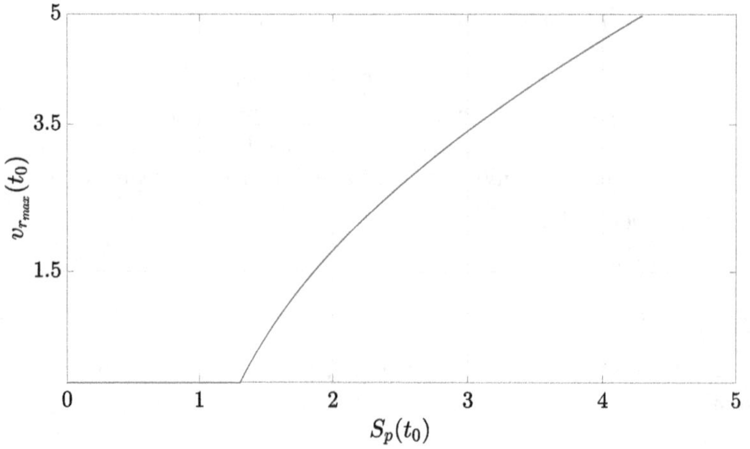

Figure 6.2 *Maximum allowed robot velocity toward the human operator in case of* $v_h(t_0) = 2$ m/s, $a_{max} = 7.5$ m/s^2, $T_r = 0.002$ s, *and* $C + Z_d + Z_r = 1.3$ m

real execution times and the real planning information to adapt online the two schedules, improving the collaboration and making it as natural as possible.

2. **The execution layer**. It is divided into the **trajectory planning** block and the **trajectory scaling** block. The first one plans and replans online robot trajectories that avoid the human operator, ensuring always a collision-free path. The second one exploits the human tracking information to adapt in real time the robot velocity along the path, making the trajectories compliant with the SSM collaborative mode defined, e.g., the magnitude of the velocity respect the constraint defined in (6.15).

The entire procedure begins at the scheduling layer, where the task assignment immediately computes the nominal schedules for both the human operator and the robot. These schedules are compliant with the precedence constraints required by the collaborative job and minimize the overall makespan. Subsequently, the two schedules are given as input to the dynamic scheduler, which has the goal to face the deviations from the nominal behaviors. It first gets the information coming from both the human tracking block, i.e., the real execution time and the actual human position inside the collaborative workspace, and the dynamic planner, i.e., task planning failures if not feasible trajectories are found or the completion of the task. Subsequently, it allocates one task at a time for each agent, optimizing the final real schedule and compensating the uncertainties that occur at runtime. The robot task is then passed to a job-dependent database, which is responsible for converting the task into a desired final configuration that the robot has to reach to complete the required task. Subsequently, this desired configuration is given as an input to the trajectory planner block to plan a collision-free trajectory. At this point, the planned trajectory is passed to the trajectory scaling block, which adapts online the magnitude of the

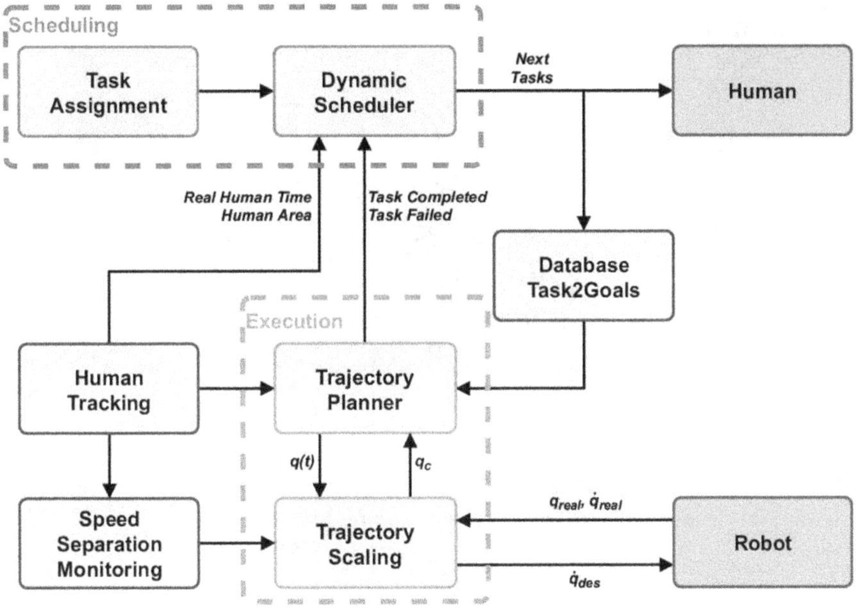

Figure 6.3 *The overall proposed framework. The blue blocks and the yellow blocks represent the scheduling layers and the execution layer, respectively. The gray blocks, instead, symbolize the supporting implemented strategies, which are necessary to process the required data. Finally, the red blocks represent the agents.*

velocity along the path, making it compliant with the safety limit imposed by the SSM and ensuring safety for the human operator. It is important to underline that the safety constraint is computed and changed at runtime based on the information coming from human tracking block. Furthermore, during the execution, the scaling block continually gives as feedback the actual status of the trajectory to the trajectory planner. This is exploited for the replanning procedure when the current trajectory becomes infeasible, i.e., the new human hinders the robot movements.

6.4.1 *Task scheduling*

The scheduling layer aims at optimally assigning and scheduling the tasks between the human and the robot, reducing as much as possible the waiting time in order to improve the collaboration. To achieve this, it executes two different steps. The first one is performed by the task assignment block, which relies on the nominal execution times to optimally assign the tasks between the two agents. This first procedure ensures the respect of the precedence constraint in order to obtain two first nominal schedules. The second step is carried out by the dynamic scheduler block, which exploits the real execution times and the actual planning information to adapt in real time the two schedules. In particular, the proposed dynamic rescheduling strategy

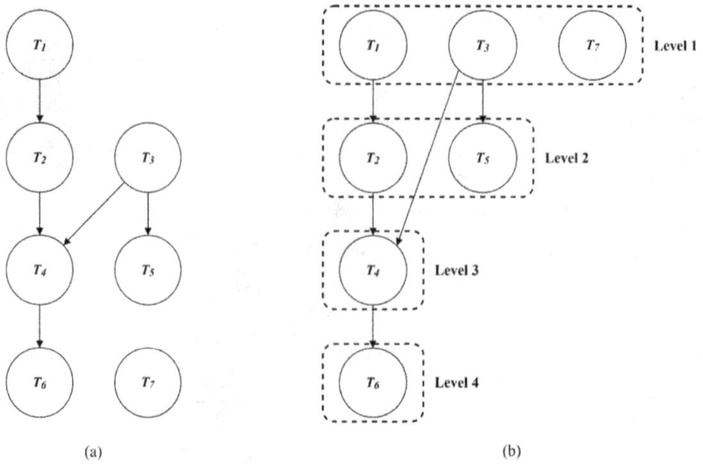

Figure 6.4 *(a) shows the directed acyclic graph of a job composed of seven tasks, while (b) shows the division into four levels*

is an extension of our previous work in Reference 27, while the communication between the two agents and the job quality constraints is not considered.

It is possible to represent each job as a directed acyclic graph $\mathcal{G} = (T, E)$, where each vertex represents the task $T_i \in T$, and each directed edge $E_{ij} \in E$ represents the precedence constraint, i.e., T_i must be executed before T_j. This is shown in Figure 6.4a. Analyzing the graph, it is possible to note that some tasks may be independent from the others. For example, in the graph, T_1 and T_3 are independent since there is not a path that goes from T_1 to T_3 and vice versa. For this reason, the graph can be redrawn in order to group all the parallel tasks into several sets called levels L_l. This is shown in Figure 6.4b. The way the tasks are grouped together is crucial for the performance of the final schedule, i.e., tasks that are not related to others can be assigned at all levels. The task assignment takes care of distributing the tasks over the levels while generating the first nominal schedule by solving the following multiobjective mixed-integer linear programming problem:

$$\min_{x,c_l} G = \sum_{l=1}^{L} \sum_{i=1}^{N} \sum_{a\in A} w_{ai} x_{ail} + \frac{1}{t_{\max}} \sum_{l=1}^{L} c_l$$

subject to

$$\sum_{l=1}^{L} \sum_{a\in A} x_{ail} = 1 \quad \forall i \in \{1, \ldots, N\} \tag{6.16}$$

$$\sum_{i=1}^{N} t_{ai} x_{ail} \le c_l \quad \forall l \in \{1, \ldots, L\}, \forall a \in A$$

$$\sum_{a\in A} \sum_{l=1}^{L} l \cdot x_{ail} < \sum_{a\in A} \sum_{l=1}^{L} l \cdot x_{ajl} \quad \forall i \to j$$

The term $w_{ai} > 0$ represents the cost required by the agent a, i.e., human or robot, to execute T_i. The calculation method of these costs is a design parameter (see, e.g., References 28 and 29), and it can take into account all the relevant metrics that the user want to consider in the task assignment procedure, e.g., electrical cost, tool wear, and agent capability. In general, the cost of the human w_{Hi} may be a combination of both the process costs and the human job quality parameters. This would allow to embed the job quality in the scheduling procedure, trying to optimize it. The Boolean variable $x_{ail} \in \{0, 1\}$ determines if T_i is assigned to the respective agent and at what level it must be executed. $t_{ai} > 0$ represents the nominal execution time of T_i on behalf of agent a, and $t_{\max}$ is the maximum task duration. Finally, $c_l > 0$ denotes the cycle time of the lth level.

The first constraint guarantees that each task is assigned to only one agent and only once. The second constraint, instead, represents the maximum parallelism criterion. In fact it ensures that, for each level, the cycle time of each agent is lower than the cycle time of the entire level c_l. Since c_l is an optimization variable, the optimization problem will choose a cycle time as low as possible, which translates into parallelizing the execution of the tasks as much as possible. Finally, the third constraint guarantees the respect of the precedence relationship imposing that the precedence tasks are assigned in an higher level. Since the levels are then executed in ascending order, this means that the precedence tasks are executed before the successor tasks.

The results of the optimization problem (6.16) are the two nominal schedules S_H and S_R, i.e., the two ordered sets of tasks that must be executed by the human and the robot, respectively, and at what level they have been assigned.

Starting from the outcome of the task assignment, the dynamic scheduler exploits the task planning information and all the human tracking data in order to adapt and, furthermore, optimize the two nominal schedules. When two humans collaborate, they immediately adapt their behavior based on what is currently happening in the working environment. If one human operator is slowing the collaboration, the other one compensates by speeding up its tasks and helping the other. Or if one human has to perform some tasks occupying a certain area, the other usually performs tasks that are not close so as not to get in the way and avoid unnecessary waste of time. The dynamic scheduler aims at reproducing this behavior, replicating this natural synergy in a team composed of a human operator and a robot, with a consequent improvement of the overall performances of the HRC.

To achieve this, the human operator is first monitored in real time while executing the task estimating the real execution time. Since the real execution time of the human operator probably differs from the nominal execution time, this information is used to evaluate the real value of the cost function G in (6.16). With the different timings introduced by the human operator, it may be convenient to reschedule some tasks for the robot, compensating the variation. To this purpose, the rescheduling procedure computes for each task T_i a penalty cost β_i, allowing to define the most suitable task for the robot at the moment. The higher this penalty cost, the lower the priority of executing the task at that time. The penalty function is computed with the following equation:

$$\beta_i = \Delta G_i + isHA_i + isFA_i + M \cdot prec_i - m \cdot succ_i \tag{6.17}$$

where ΔG_i represents the variation of the cost function in (6.16) that would occur if the task T_i is performed by the robot at this moment. $isHA_i$ and $isFA_i$ are two Boolean variables used to exploit the planning information inside the dynamic rescheduling strategy. When the human operator is occupying the area of the task T_i, it is very imaginable that the planner will not be able to find a feasible trajectory for the robot, causing a failure. Or, the closeness of the human operator would be reflected in a large reduction in the robot velocity, with a consequent dropping of the performances. For these reasons, it would be better to execute other tasks, setting the variable $isHA_i$ equal to one and penalizing T_i. Another situation occurs when the planner is not capable of finding a feasible trajectory, and the task execution has failed. Indeed, it could happen that, even if the human operator is not in the same area where the robot has to execute the task T_i, its position hinders the robot, making it impossible for the execution layer to find a collision-free trajectory. This is handled by setting the variable $isFA_i$ equal to one, penalizing all the tasks that have to be executed in the same area. Finally, the variables $prec_i$ and $succ_i$ represent the number of precedences and successor tasks that have not yet been concluded, respectively. These variables are multiplied by a large constant M and a small constant m, in order to penalize tasks whose precedences have not yet been executed, e.g., β_i assumes a large value and favors tasks with more successors, giving the possibility of executing more tasks in the next steps.

The rescheduling procedure that exploits this penalty cost is executed by the dynamic scheduler block. Specifically, its implementation is represented by the pseudocode reported in Algorithm 1. The algorithm requires as an input the two nominal task schedules S_H and S_R (Line 1). First, it sets to false the two variables End_H and End_R (Line 2). These variables are used to identify when the respective agent has performed the assigned task and when it is ready to start a new one. Subsequently, the algorithm initializes the variable $Fail$, which is used to identify when the execution layer has failed in finding a feasible trajectory (Line 3). At this point, the overall scheduling procedure is initialized at the first level, and the dynamic scheduler starts a loop to iterate through all the levels of the schedule (Lines 4–5). Inside the loop, as the first step, the algorithm checks if the human operator is free, and if this is the case, it assigns him a new task (Line 6). Subsequently, it makes the most of the human monitoring data to obtain both the real execution time of the human operator and its occupancy area (Line 8). These information are then exploited to update the cost function G, obtaining the estimation of the real value (Line 9). The procedure **monitorH**() is out of the scope of this work, and it can be implemented using several strategies already available in the literature (see, e.g., References 24–26). At this point, the algorithm checks if it is necessary to update the failing area and defining k as the number of tasks that has been already completed computes

$$\mathrm{B}_R = \{(T_{k+1}, \beta_{k+1}), \ldots, (T_N, \beta_N)\}$$

namely the tuple that contains the tasks that have not yet completed with their corresponding penalty costs (Lines 10–12). It is worth noting that when the number of the remaining tasks is large, evaluating β_i for all of them could be a very demanding procedure. To address this, it is possible to take advantage of how the job is represented by limiting the computation to only the tasks that were originally scheduled in the actual level and in a future desired horizon. Moreover, it also important to underline that β_i

is calculated for all remaining tasks, whether they have been initially scheduled to the robot or the human operator. As a result, the robot could perform a task that was initially assigned to the human operator, if necessary. Subsequently, if the robot is available, the dynamic scheduler assigns the cheapest task to the robot, updating the schedules S_R and S_H, and reinitializes the variable $Fail$ to false (Lines 14–15). The **assign**() function must also check that the same task is not assigned to both agents and that the next robot task is not too expensive, e.g., it must avoid assigning tasks whose precedences have not yet been completed or tasks that cannot be performed by the robot. This procedure can be implemented using standard sorting algorithm, e.g., knapsack algorithm [30]. Finally, the algorithm checks if the human operator and the robot have concluded their tasks or if the planner has failed (Lines 17–18) and if all the tasks in the actual level have been concluded, i.e., the two agents have no tasks (Lines 19–20).

Algorithm 1: DynamicScheduler()

1: **Require:** S_H,S_R
2: $End_H, End_R \leftarrow false$
3: $Fail \leftarrow false$
4: $l \leftarrow 1$
5: **while** $l \leq L$ **do**
6: **if** End_H **then** $T_H \leftarrow$ **next**(S_H)
7: **end if**
8: $(t_{real}, HA) \leftarrow$ **monitorH**()
9: **updateG**(t_{real})
10: **if** $Fail$ **then** $FA \leftarrow setFA(T_R)$
11: **end if**
12: $\mathrm{B}_R \leftarrow$ **computeBeta**(S_R, HA, FA)
13: **if** End_R **or** $Fail$ **then**
14:   $T_R \leftarrow$ **assign**$(\mathrm{B}_R, T_H, S_R, S_H)$
15:   $Fail \leftarrow false$
16: **end if**
17: $End_H \leftarrow$ **checkEndH**()
18: $(End_R, Fail) \leftarrow$ **checkPlanner**()
19: **if** $T_H = \emptyset$ **and** $T_R = \emptyset$ **then**
20:   $l \leftarrow l + 1$
21: **end if**
22: **end while**

Once the dynamic scheduler assigns the next robot task T_R, this task must be converted into a desired final configuration in order to allow the planning layer to find a feasible trajectory. This conversion is performed exploiting a case-dependent database, which, for each task T_i, returns the desired final configuration q_f. It is worth noting that the scheduling layer could be easily extended by integrating the job quality constraints over the entire working shift and the communication between the two agents, as shown in the previous work [27]. Unlike this previous work, however, the rescheduling strategy proposed in this chapter always find the best task for the robot and allows the robot to perform and anticipate tasks that were previously assigned to the human operator, improving the performances of the HRC.

6.4.2 Task execution

The planning layer aims at allowing the robot to execute the assigned tasks while ensuring safety for the human operator. To achieve this, it executes two different steps. The first one is performed by the trajectory planning block, which plans a trajectory that is collision-free and that the robot could ideally execute at maximum speed. The second, instead, is performed by the trajectory scaling block, which adapts at runtime only the magnitude of the velocity along the path, ensuring that the trajectory is compliant with the SSM collaborative mode (see section 6.3). The planning layer is based on the previous work in Reference 31, of which it is an extension as it is integrated into a very complex architecture.

The dynamic planning procedure is executed by the trajectory planner block. Specifically, its implementation is represented by the pseudocode reported in Algorithm 2.

Algorithm 2: TrajcetoryPlanner()

```
1: Require: q_i, q_f
2: (q_d(·), Fail) ← plan(q_i, q_f)
3: send(q_d(·))
4: q_c ← q_i
5: while q_c ≠ q_f and not Fail do
6:     if not_feasible(qc, h_P) then
7:         (q_d(·), Fail) ← replan(q_d(·), q_feas, q_f)
8:         send(q_d(·))
9:         break
10:    end if
11:    update_q_c()
12: end while
13: if q_c = q_f then
14:    End_R ← true
15: else
16:    End_R ← false
17: end if
18: updateScheduling(End_R, Fail)
```

The algorithm requires as input the initial configuration q_i and the desired final configuration q_f coming out from the case-dependent database (Line 1). First, it computes a collision-free trajectory $q_d(\cdot)$, taking into account all the obstacles in the working area and the kinodynamic constraints of the robot, i.e., a trajectory that the robot could ideally execute at maximum speed (Line 2). If the trajectory planner is not capable of finding any feasible trajectories, the result is an empty trajectory, and the variable *Fail* is set to false. The function **plan()** can be implemented using planning algorithm suitable for robotic applications that are already available in the literature (see, e.g., References 32–34). Subsequently, the algorithm sends the entire trajectory to the trajectory scaling block, and it initializes the current trajectory state q_c equal to the initial configuration q_i (Lines 3–4). At this point, the trajectory planner starts a loop until the trajectory is completed or some failures occur. Inside the loop, the algorithm continuously checks for the feasibility of the trajectory (Line 6). Since the environment is highly dynamic, i.e., the human operator moves inside the working area, it is very likely that the planned trajectory become soon infeasible, and a new one is required. To speed up the process and avoid useless replanning, the checking procedure is performed only on a portion of the trajectory, i.e., over a horizon h_P that starts from the actual robot configuration q_c. If the checking procedure finds at least one infeasible configuration, a new collision-free trajectory is planned through the function **replan**() (Line 7). The **replan**() function takes care of planning the new collision-free trajectory that goes from a desired configuration, in this case, the least feasible robot configuration, q_{feas}, to the desired final configuration. Moreover, to ensure continuity for the trajectory scaling block, the **replan**() function is also responsible for merging the new trajectory with the feasible portion of the previous one. The overall merged trajectory can then be sent to the trajectory scaling block (Line 8). Subsequently, the algorithm exploits the available data to update the actual state (Line 11). When the trajectory has been completed or no feasible trajectories have been found, i.e., the replanning procedure has failed, the algorithm exits the while loop and sets the variable End_R to true. Finally, the two variables End_R and $Fail$ are sent to the dynamic scheduler to enrich the scheduling procedure with the planning information and improve the collaboration (see Algorithm 1, Line 18).

Once the collision-free trajectory has been planned by the trajectory planner, the trajectory scaling block generates the control input for the robot. The aim of this block is to ensure that the robot behavior is compliant with the safety constraint imposed by the ISO/TS 15066 standard (6.15). To this purpose, it regulates and scales online only the magnitude of the velocity $\dot{s}$ of the trajectory, without modifying the planned path that is collision-free.

First, the trajectory scaling applies the path-velocity decomposition, as shown in (6.1) and (6.2), to the trajectory. Subsequently, it finds the maximum admissible velocity by solving online the following optimization problem:

$$
\begin{aligned}
&\min -\alpha \\
&\text{subject to} \\
&J_{r_i}(q)q'_d(s)\dot{s}\alpha \le V_{\max_i} \quad \forall i \in \{1,\ldots,n\} \\
&\dot{q}_{\min} \le q'_d(s)\dot{s}\alpha \le \dot{q}_{\max} \\
&\ddot{q}_{\min} \le \frac{q'_d(s)\dot{s}\alpha - \dot{q}}{T_r} \le \ddot{q}_{\max} \\
&0 \le \alpha \le 1
\end{aligned}
\tag{6.18}
$$

$\alpha \in [0, 1]$ is the optimization variable and represents the scaling factor. $J_{r_i}(q) \in \mathbb{R}^{1\times n}$ is a *modified Jacobian* that takes into account only the scalar velocity toward the human operator of the ith link. This modified version of the Jacobian method allows to ensure that the velocity is compliant with the ISO/TS 15066 standard (6.15) since the safety standards regulate only the velocity that reduces the human-robot distance, i.e., the velocity toward the human. $V_{\max_i} \in \mathbb{R}$ is the velocity limit imposed by the ISO/TS 15066 standard for the ith robot link. $\dot{q}_{\min} \in \mathbb{R}^n$ and $\dot{q}_{\max} \in \mathbb{R}^n$ are the joint velocity lower bounds and the joint velocity upper bounds, respectively. While $\ddot{q}_{\min} \in \mathbb{R}^n$ and $\ddot{q}_{\max} \in \mathbb{R}^n$ are the acceleration limits. $\dot{q} \in \mathbb{R}^n$ is the actual robot velocity, and T_r is the robot execution time.

The modified Jacobian $J_{r_i}(q)$ is expressed as

$$J_{r_i}(q) = \vec{n}_i^T \begin{bmatrix} J_i(q) & \bar{0} \end{bmatrix} \tag{6.19}$$

where $\vec{n}_i = \{n_{x_i}, n_{y_i}, n_{z_i}, 0, 0, 0\}$ is the unit vector that represents the ith robot–human direction, i.e., the direction that goes from the ith robot link to the human operator. This direction can be computed exploiting several strategies already available in the literature, e.g., it can be found representing both the robot and the human links as capsules and computing the minimum distance [35] between all the capsules. $J_i(q) \in \mathbb{R}^{6\times i}$ is the ith Jacobian, i.e., the Jacobian matrix that relates the first i joint velocities to the linear and angular velocities of the ith link, and $\bar{0} \in \mathbb{R}^{6\times(n-i)}$ is a matrix with all zero elements necessary for matching the dimensions.

It is worth noting that the optimization problem (6.18) is a convex problem and, therefore, computationally cheap. Indeed, the only factor that affects the convergence is the problem dimension, i.e., the number of joints and links, which is low for a general HRC application. Remembering the properties of the convex optimization problems, the solution found by the solver is always the global minimum of the cost function, i.e., the maximum admissible scaling factor. Moreover, the problem has always a feasible solution. When the human operator is very far from the robot, the robot is allowed to follow the planned trajectory, i.e., $\alpha = 1$, which is the maximum velocity. When the distance between the human operator and the robot decreases, the safety standard requires to reduce the velocity until, in the worst case, stopping the robot. This is ensured by the solution $\alpha = 0$.

The output of the trajectory scaling is then used to control the robot, sending the optimal safe velocity:

$$\dot{q} = q_d'(s)\dot{s}\alpha \tag{6.20}$$

And the curvilinear abscissa s is updated accordingly:

$$s_{\text{new}} = s + \dot{s}\alpha T_r \tag{6.21}$$

Finally, the new robot configuration that is given to the dynamic planner for the replanning procedure (see Algorithm 2, Line 11) is equal to:

$$q_c = q_d(s_{\text{new}}) \tag{6.22}$$

Reducing the robot velocity is not always the best solution in term of performances. Sometimes it may be better to change the overall path, finding a trajectory that is maybe longer but faster, e.g., a trajectory that is far away from the human operator. This can be achieved by implementing a feedback signal procedure, as demonstrated in Reference 31.

6.5 Experiments

The proposed two-layered architecture has been experimentally validated in a custom HRC scenario. During the experiments, the human operator has to cooperate with the UR10e, a 6-DoF collaborative robot. To monitor the human task execution, five OptiTrack Primex cameras with the OptiTrack Motive software* have been exploited. To evaluate the remaining task time, the open-end dynamic time warping (OE-DTW) algorithm, already available in literature [26], has been implemented to both operator wrists. Due to its low reliability, the OE-DTW is supported with a simple HMI that allows the human operator to confirm the completion of the task. The complete setup for the experiments is shown in Figure 6.5.

All the software components were developed using ROS Melodic Morenia metaoperating system, and they ran on a Intel(R) Core(TM) i7-10510U with Ubuntu 18.04. The task scheduling layer exploits Python-MIP†, a collection of Python tools for the modeling and solution of mixed-integer linear programs, and the Gurobi solver‡ to compute the first optimal nominal schedules. The task execution layer, instead, relies on MoveIt motion planning framework [36] with the RRT-connect algorithm [37] to plan the collision-free trajectory. The optimization problem (6.18) in the scaling block, instead, is automatically solved with the C code generated by the fmincon MATLAB® function. Concerning the frequencies, the communication with the robot works at 500 Hz while the optimization problem converges in 1 ms. The OptiTrack, instead, works at a frequency of 240 Hz.

Table 6.1 reports a detailed description of all the tasks composing the collaborative job, while the data required by the scheduling layer (see section 6.4.1) are shown in Table 6.2. The column $\boldsymbol{i \rightarrow j}$ represents the precedence constraint, e.g., tasks T_6 and T_7 must be executed before tasks T_4 and T_5. The nominal durations have been defined by performing the tasks multiple times and computing the average time. The intrinsic costs, instead, have been defined using our experience and knowledge in the collaborative job. It is worth noting that $T_1 = T_2 = T_3$, e.g., there are three tasks that are the same, and for these tasks, w_{Ri} is very high, this is because the robot is not capable of executing the USB packaging action. Table 6.3, instead, reports the data used in the scaling optimization problem (6.18).

Exploiting these data, the task assignment block solves the optimization problem (6.16) in 100 ms, whose results are the following optimal nominal schedules:

*https://www.optitrack.com/.
†https://www.python-mip.com/.
‡http://www.gurobi.com/.

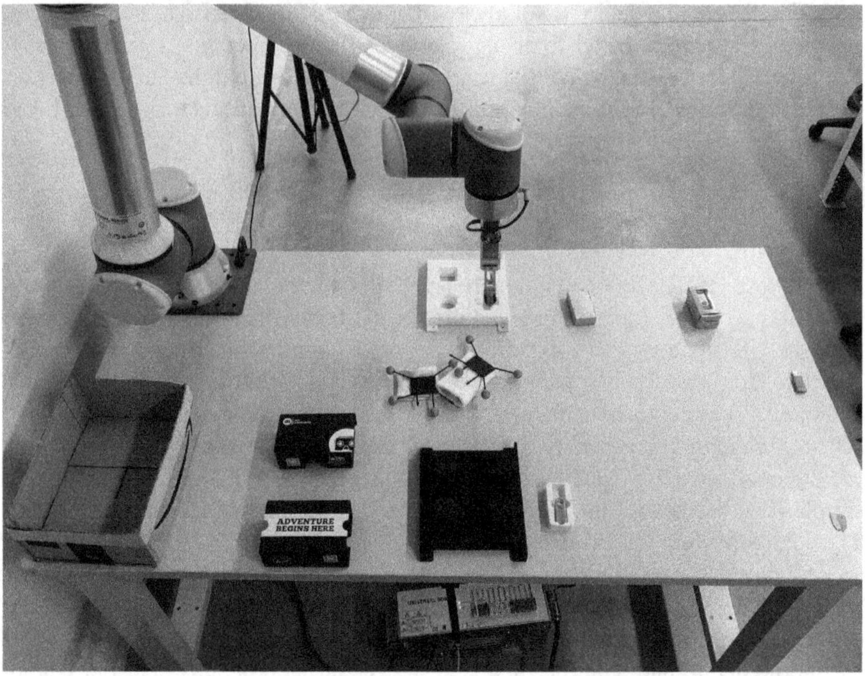

Figure 6.5 *Setup of the experiments. The image shows all the equipment used during the experiment. In practical, it is possible to notice two wristbands used to track both human wrists with the OptiTrack.*

- $S_H = \{[2,3,8],[1]\}$
- $S_R = \{[6,7,9],[4,5]\}$

Starting from the output of the task assignment, the dynamic scheduler is then tested into three different scenarios: when the human operator is very slow in

Table 6.1 *Tasks description*

Task index	Description
1	Packaging USB key.
2	Packaging USB key.
3	Packaging USB key.
4	Pick-and-place circular shape.
5	Pick-and-place cross shape.
6	Pick-and-place square shape.
7	Pick-and-place U shape.
8	Pick-and-place weight.
9	Pick-and-place weight.

Table 6.2 Task assignment data

Task index	w_{Ri}	t_{Ri}	w_{Hi}	t_{Hi}	$i \rightarrow j$	Area
1	1,000	–	0.4	25	–	1
2	1,000	–	0.4	25	–	1
3	1,000	–	0.4	25	–	1
4	0.1	12	0.4	15	–	2
5	0.1	12	0.4	15	–	2
6	0.1	12	0.4	15	4, 5	2
7	0.1	12	0.4	15	4, 5	2
8	0.5	25	0.8	10	–	3
9	0.5	25	0.8	10	–	3

Table 6.3 Planning data description

Data	Value
a_{max}	−3.71 m/s^2
T_r	0.002 s
C	0.21 m
Z_d	0.0 m
Z_r	0.01 m
$\dot{q}_{max} = -\dot{q}_{min}$	3.14 rad/s
$\ddot{q}_{max} = -\ddot{q}_{min}$	6.28 rad/s^2

executing tasks, the human hinders the robot during the collaboration, the human is inside the same working area of the next robot task. In particular, the first scenario is performed twice: first with a baseline architecture and, later, with the proposed architecture. The baseline architecture is a simplified architecture that does not implement any rescheduling strategy and, therefore, waits for the completion of all the tasks of each level. The other two scenarios are necessary to show how the proposed architecture dynamically adapts the schedule both in the case of planning failures and already occupied working area, validating the overall architecture. A video showing all the experiments is attached, and it can be found in the supplementary material.

A detailed comparison of the timing of the experiments in the first scenario is reported in Figure 6.6. Figure 6.6a shows the ideal optimal nominal schedule computed by the task assignment block exploiting the nominal durations, i.e., the schedule that would be obtained if there were no deviations from the nominal behavior. In Figure 6.6b, instead, the real schedule obtained with the baseline architecture is reported. Initially, both the human operator and the robot start to follow the ideal nominal schedules. However, according to the proposed scenario, the human operator is slow in executing the assigned tasks, and in the meanwhile, the robot is able

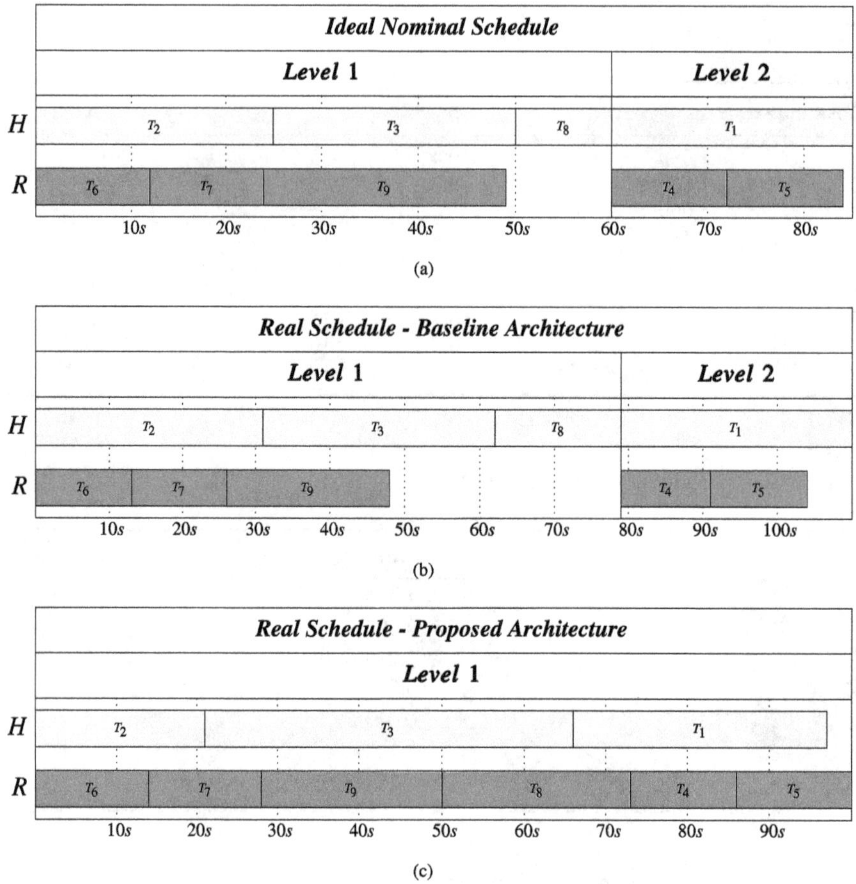

Figure 6.6 *Gantt chart of the schedules for the first scenario. (a) shows the ideal schedule that would be obtained with nominal durations. (b) represents the real schedule resulting with the baseline architecture, while (c) is the real schedule obtained with the proposed architecture.*

to conclude the first three tasks, i.e., tasks T_6, T_7, and T_9. Since no rescheduling procedure is allowed, the robot stops and waits for the human to conclude all the tasks in the first level of the schedule. After the completion of task T_8, the schedule passes to the second level, and two agents start again to execute their tasks, resuming the expected behavior. The total duration of the schedule obtained with the baseline architecture is $t_{tot} = 110$ s.

The proposed architecture, instead, relies on the rescheduling strategy, discussed in section 6.4.1, to dynamically adapt the schedules, compensating for the deviations from the nominal behaviors and improving the performances of the HRC. The resulting schedule is shown in Figure 6.6c. Unlike what happened with the baseline architecture, during the execution, the dynamic scheduler exploits both the OE-DTW algorithms to compute the variation of the cost function G in (6.16) caused by the human delay and the skeleton tracking information to update the tuple of the penalty costs B_R (see Algorithm 1). After the robot has concluded all its tasks in the first level, the delay introduced by the human operator would cause a drop of the performances, i.e., the parallelism criterion in the first level is not achieved. For this reason, the dynamic scheduler returns that is better to reschedule T_8 to the robot, helping the human and making the collaboration more efficient. At this point, the robot completes T_8, and both agents are able to conclude the remaining schedules. It is worth noting that, in the graph, the final schedule is composed of only one level. This is because, according to Algorithm 1, the dynamic scheduler anticipated all the tasks to the first level, making the others empty. The total duration of the schedule obtained with the proposed architecture is $t_{tot} = 100$ s.

6.6 Discussion

The experiments carried out show that the proposed framework is able to achieve a good scheduling procedure, significantly improving the performance compared with a static architecture. In fact, the total makespan was reduced by 9.1%. Besides the improvement in the total makespan, a great improvement introduced by the proposed architecture can be seen in the robot behavior. With the dynamic rescheduling strategy proposed in this work, the robot is able to execute a task instead of the human, helping in the job. This is exactly the behavior that would occur in the case of a collaboration between two human operators. With the baseline architecture, instead, the robot has to wait until the entire level is concluded, without compensating any human delay. Moreover, the proposed architecture is capable of assigning a new task if the human hinders the robot, avoiding useless waiting time. While the baseline architecture continues to reassign the same task waiting for the operator to move. This is possible, thanks to the mutual integration with a proper safe planner strategy, proving that the communication between the two layers is desirable.

The proposed approach, however, has some limitations that are worth discussing. First of all, to be compliant with the ISO/TS 15066 also, the monitoring unit should be safe. To this aim, the use of the OptiTrack is necessary to simulate and substitute a sensing system that is certified for safety critical application. However, it cannot be directly used in real industrial settings for applications, where safety is involved. Moreover, certified sensing systems are usually robust to the occlusion that may arise in a real HRC application. This is why, these possible failures in the detection are not considered in this work. However, when certified systems are not available, it is possible to ensure safety by detecting such occlusions and applying a strategy like in Reference 38, where the robot behavior is made compliant during the execution of a task.

Another issue that should be considered when dealing with the real implementation of the framework regards the numerical integration of the curvilinear abscissa $s(t)$ (6.21). Since the control frequency of the robot is 500 Hz, the integration error is very small and, in some cases, may be neglected, as shown in Figure 6.7. However, when the required precision is very high, e.g., pick-and-place application, this error should be corrected. This can be done by slightly modifying the optimization problem in (6.18), where $q_d'(s)\dot{s}$ is substituted with

$$\dot{q}_{\text{des}} = q_d'(s)\dot{s} + K_p(q_d(s) - q_{\text{real}}) + K_v(q_d'(s)\dot{s} - \dot{q}_{\text{real}}) \tag{6.23}$$

where $K_p \in \mathbb{R}^{n\times n}$ and $K_v \in \mathbb{R}^{n\times n}$ are the diagonal matrices of the position and velocity error, respectively. The plots of the resulting behavior with this modification are shown in Figure 6.8.

6.7 Conclusion

In this paper, we proposed a two-layered dynamic framework that mutually integrates both the task scheduling and the task execution procedures. The framework relies on a human monitoring strategy in order to obtain both the estimation of the real execution time and the actual human position. These data are then exploited by the two layers to reschedule the tasks and replan the trajectories, improving the performances of the collaboration. The overall architecture has been tested in a custom HRC assembly scenario with promising results. The robot helped the human

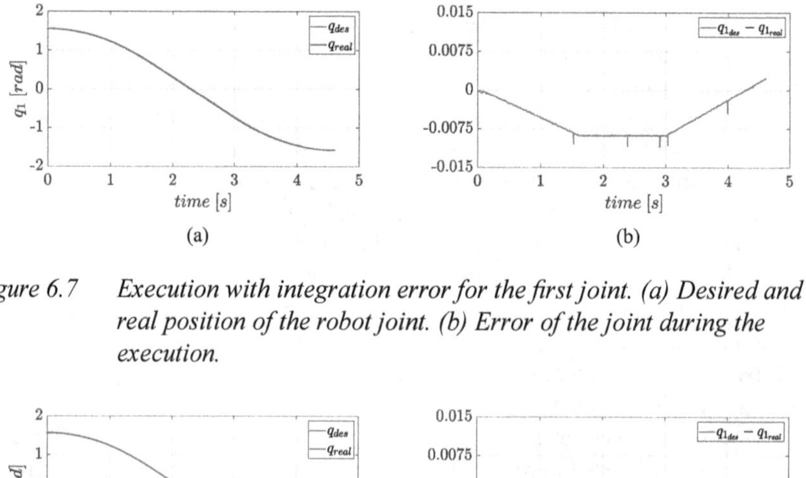

Figure 6.7 Execution with integration error for the first joint. (a) Desired and real position of the robot joint. (b) Error of the joint during the execution.

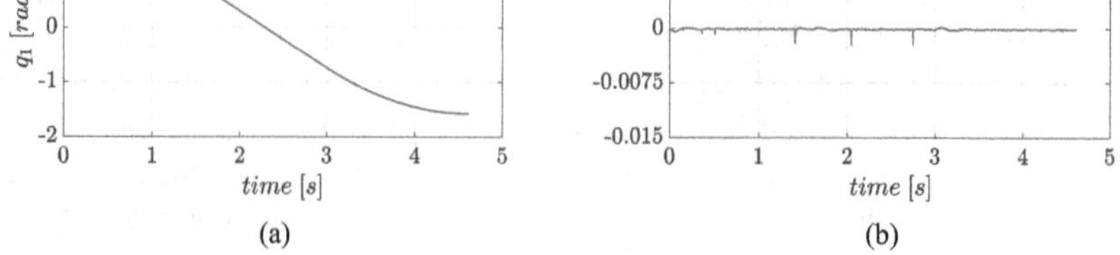

Figure 6.8 Execution with K_p = diag(10) and K_v = diag(0) for the first joint. (a) Desired and real position of the robot joint. (b) Error of the joint during the execution.

operator when it was late. Or, when necessary, the robot task was rescheduled online in order to avoid idle times or speed reduction.

Future work aims at validating the proposed architecture in a real industrial scenario. This would allow to perform also a user study in order to get the feedback from inexpert users and further validate the architecture. Moreover, instead of performing a task assigned to the human, it might be interesting to investigate how the robot can help the human speed up the execution of the actual assigned task, e.g., handing some objects. Finally, HMI could be replaced with a voice control architecture, making the human-robot communication more like the natural human communication, and the OE-DTW could be improved or substituted with a more reliable strategy.

References

[1] Villani V., Pini F., Leali F., Secchi C. 'Survey on human-robot collaboration in industrial settings: safety, intuitive interfaces and applications'. *Mechatronics*. 2018, vol. 55, pp. 248–66.

[2] *Robots and robotic devices–safety requirements for industrial robots–part 1: robots*. Geneva, CH: International Organization for Standardization; 2011.

[3] *Robots and robotic devices–safety requirements for industrial robots–part 2: robot systems and integration*. Geneva, CH: International Organization for Standardization; 2011.

[4] *Robots and robotic devices–collaborative robots*. Geneva, CH: International Organization for Standardization; 2016.

[5] Zanchettin A.M., Ceriani N.M., Rocco P., Ding H., Matthias B. 'Safety in human-robot collaborative manufacturing environments: metrics and control'. *IEEE Transactions on Automation Science and Engineering*. 2016, vol. 13(2), pp. 882–93.

[6] Levratti A., Riggio G., Fantuzzi C., De Vuono A., Secchi C. 'TIREBOT: a collaborative robot for the tire workshop'. *Robotics and Computer-Integrated Manufacturing*. 2019, vol. 57, pp. 129–37.

[7] Ferraguti F., Bertuletti M., Landi C.T., Bonfe M., Fantuzzi C., Secchi C. 'A control barrier function approach for maximizing performance while fulfilling to ISO/TS 15066 regulations'. *IEEE Robotics and Automation Letters*. 2020, vol. 5(4), pp. 5921–28.

[8] Tsarouchi P., Matthaiakis A.-S., Makris S., Chryssolouris G. 'On a human-robot collaboration in an assembly cell'. *International Journal of Computer Integrated Manufacturing*. 2017, vol. 30(6), pp. 580–89.

[9] Blum C., Miralles C. 'On solving the assembly line worker assignment and balancing problem via beam search'. *Computers & Operations Research*. 2011, vol. 38(1), pp. 328–39.

[10] Xu K., Fei R., He D. 'A Tabu-search algorithm for scheduling jobs with precedence constraints on parallel machines'. *2018 13th IEEE Conference on Industrial Electronics and Applications (ICIEA)*; IEEE, 2018. pp. 2774–81.

[11] Sabattini L., Digani V., Lucchi M., Secchi C., Fantuzzi C. 'Mission assignment for multi-vehicle systems in industrial environments'. *IFAC-PapersOnLine*. 2015, vol. 48(19), pp. 268–73.

[12] Michalos G., Spiliotopoulos J., Makris S., Chryssolouris G. 'A method for planning human robot shared tasks'. *CIRP Journal of Manufacturing Science and Technology*. 2018, vol. 22, pp. 76–90.

[13] Li K., Liu Q., Xu W., Liu J., Zhou Z., Feng H. 'Sequence planning considering human fatigue for human-robot collaboration in disassembly'. *Procedia CIRP*. 2019, vol. 83, pp. 95–104.

[14] Ayough A., Zandieh M., Farhadi F. 'Balancing, sequencing, and job rotation scheduling of a u-shaped lean cell with dynamic operator performance'. *Computers & Industrial Engineering*. 2020, vol. 143, p. 106363.

[15] Rahman S.M.M., Wang Y. 'Mutual trust-based subtask allocation for human-robot collaboration in flexible lightweight assembly in manufacturing'. *Mechatronics*. 2018, vol. 54, pp. 94–109.

[16] Nikolakis N., Kousi N., Michalos G., Makris S. 'Dynamic scheduling of shared human-robot manufacturing operations'. *Procedia CIRP*. 2018, vol. 72, pp. 9–14.

[17] Johannsmeier L., Haddadin S. 'A hierarchical human-robot interaction-planning framework for task allocation in collaborative industrial assembly processes'. *IEEE Robotics and Automation Letters*. 2017, vol. 2(1), pp. 41–48.

[18] Casalino A., Zanchettin A.M., Piroddi L., Rocco P. 'Optimal scheduling of human-robot collaborative assembly operations with time petri nets'. *IEEE Transactions on Automation Science and Engineering*. 2021, vol. 18(1), pp. 70–84.

[19] Pupa A., Landi C.T., Bertolani M., Secchi C. 'A dynamic architecture for task assignment and scheduling for collaborative robotic cells' in *International workshop on human-friendly robotics*. Springer; 2020. pp. 74–88.

[20] Pupa A., Secchi C. 'A safety-aware architecture for task scheduling and execution for human-robot collaboration'. *2021 IEEE/RSJ International Conference on Intelligent Robots and Systems (IROS)*; Prague, Czech Republic, 2021. pp. 1895–902.

[21] Moon S., Park Y., Ko D.W., Suh I.H. 'Multiple kinect sensor fusion for human skeleton tracking using Kalman filtering'. *International Journal of Advanced Robotic Systems*. 2016, vol. 13(2), p. 65.

[22] Kofman J., Luu T.J., Verma S. 'Teleoperation of a robot manipulator using a vision-based human-robot interface'. *IEEE Transactions on Industrial Electronics*. 2005, vol. 52(5), pp. 1206–19.

[23] Fan J., Xu W., Wu Y., Gong Y. 'Human tracking using convolutional neural networks'. *IEEE Transactions on Neural Networks*. 2010, vol. 21(10), pp. 1610–23.

[24] Vo N.N., Bobick AF. 'From stochastic grammar to bayes network: probabilistic parsing of complex activity'. *2014 IEEE Conference on Computer*

Vision and Pattern Recognition (CVPR); Columbus, OH, USA, 2014. pp. 2641–48.

[25] Maeda G., Ewerton M., Neumann G., Lioutikov R., Peters J. 'Phase estimation for fast action recognition and trajectory generation in human-robot collaboration'. *The International Journal of Robotics Research*. 2017, vol. 36(13–14), pp. 1579–94.

[26] Maderna R., Lanfredini P., Zanchettin A.M., Rocco P. 'Real-time monitoring of human task advancement'. *2019 IEEE/RSJ International Conference on Intelligent Robots and Systems (IROS)*; Macau, China, IEEE, 2019. pp. 433–40.

[27] Pupa A., Van Dijk W., Secchi C. 'A human-centered dynamic scheduling architecture for collaborative application'. *IEEE Robotics and Automation Letters*. 2021, vol. 6(3), pp. 4736–43.

[28] Lamon E., De Franco A., Peternel L., Ajoudani A. 'A capability-aware role allocation approach to industrial assembly tasks'. *IEEE Robotics and Automation Letters*. 2019, vol. 4(4), pp. 3378–85.

[29] Liu Z., Wang X., Cai Y., *et al.* 'Dynamic risk assessment and active response strategy for industrial human-robot collaboration'. *Computers & Industrial Engineering*. 2020, vol. 141, p. 106302.

[30] Dawande M., Kalagnanam J., Keskinocak P., Salman F.S., Ravi R. 'Approximation algorithms for the multiple knapsack problem with assignment restrictions'. *Journal of Combinatorial Optimization*. 2000, vol. 4(2), pp. 171–86.

[31] Pupa A., Arrfou M., Andreoni G., Secchi C. 'A safety-aware kinodynamic architecture for human-robot collaboration'. *IEEE Robotics and Automation Letters*. 2021, vol. 6(3), pp. 4465–71.

[32] LaValle S.M. *Rapidly-Exploring Random Trees: A New Tool for Path Planning*. The annual research report, 1998.

[33] Jaillet L., Simeon T. 'A PRM-based motion planner for dynamically changing environments'. *IEEE/RSJ International Conference on Intelligent Robots and Systems (IROS) (IEEE Cat. No.04CH37566)*; 2004. pp. 1606–11.

[34] Ratliff N., Zucker M., Bagnell J.A., Srinivasa S. 'CHOMP: gradient optimization techniques for efficient motion planning'. *2009 IEEE International Conference on Robotics and Automation (ICRA)*; Kobe, Japan, IEEE, 2009. pp. 489–94.

[35] Ferraguti F., Talignani Landi C., Costi S., *et al.* 'Safety barrier functions and multi-camera tracking for human-robot shared environment'. *Robotics and Autonomous Systems*. 2020, vol. 124, 103388.

[36] Coleman D., Sucan I., Chitta S, *et al.* 'Reducing the barrier to entry of complex robotic software: a moveit! case study'. *arXiv Preprint arXiv:14043785*. 2014.

[37] Kuffner J.J., LaValle S.M. 'RRT-connect: an efficient approach to single-query path planning'. *Proceedings 2000 ICRA. Millennium Conference. IEEE International Conference on Robotics and Automation. Symposia Proceedings (Cat. No.00CH37065)*; San Francisco, CA, USA, IEEE, 2000. pp. 995–1001.

[38] Udai A.D., Hayat A.A., Saha SK. 'Parallel active/passive force control of industrial robots with joint compliance'. *2014 IEEE/RSJ International Conference on Intelligent Robots and Systems (IROS 2014)*; Chicago, IL, USA, IEEE, 2014. pp. 4511–16.

Chapter 7

Tactile sensing for physical human-robot interaction in shared payload tasks

Francesco Grella[1], Keerthi Sagar[1], Roberto Canale[1], Michal Jilich[1], Alessandro Albini[2], Giorgio Cannata[1], and Matteo Zoppi[1]

Physical human-robot interaction (pHRI) is a major challenge in robotics, especially when related to collaborative operations involving the handling of heavy payloads by potentially dangerous robots. In this chapter, we present a sensing architecture designed for *shared payload tasks* and implemented to accomplish collaborative car windshield manipulation. The pHRI problem is tackled by implementing *voluntary contact recognition* leveraging data-driven methodologies, ensuring that robot motion is enabled only when the operator is deliberately interacting with the robot. Test bench experiments are eventually presented to show the functionality of the handles along with the integration of the devices on a vacuum gripper for implementation on an industrial robotic manipulator.

7.1 Introduction

Within the context of *Industry 4.0*, flexible robotic technologies are an increasing necessity since robots are required to be more than simple assembly chain elements with predefined tasks. Manipulators are expected to exhibit compliance with respect to the environment in order to adapt and properly tune their behavior during task execution [1, 2]. Moreover, human operators must be taken into account as a fundamental element of robotic applications since they represent external agents with their own specific dynamics. Physical human-robot interaction (pHRI) in industrial contexts is a matter of concern under social and safety aspects, as addressed in occupational safety literature such as Reference 3, showing the fact that novel robotic technologies are required to implement and promote safe and proficient collaboration between man and machine. Along with this introduction,

[1]Universitá degli Studi di Genova, Genova, Italy
[2]Oxford Robotics Institute, University of Oxford, England

the authors aim to point out the difference between two apparently similar concepts: *cooperation* and *collaboration*. The substantial difference mainly regards the nature of the interaction. Here, the authors are relating *cooperation* to a wide spectrum of situations in which an human interacts with the robot asynchronously, for instance during a static cooperative assembly task in which the robot waits for the human to perform the next action (i.e., passing the next part or a tool).

On the other hand, *collaboration* is identified as a continuous interaction between the human and the machine, highlighting the concept of the operator and robot as a unified system so that a collaborative task cannot be accomplished asynchronously but requires active simultaneous participation from both involved agents. Thus, this chapter is centered on a problem related to *continuous physical interaction* and aims to thoroughly describe our proposed solution.

The main concepts of the *pHRI* field are exhaustively reviewed in References 4 and 5. Problems related to *pHRI* scenarios are well-known in robotics and artificial intelligence literature but are mostly restricted to Cobots applications [6–8]. Such systems are developed specifically for interaction tasks, and major design efforts are related to compliance and safety measures. Typically Cobots are designed and built with compliant actuators like series elastic actuators [9] or variable stiffness actuators [10]. They are also provided with an embodied sensing apparatus compliant to modern safety standards and capable to prevent collisions within the *peripersonal space* of the robot, leading to an immediate motion stop when necessary. The novel paradigms defined within the context of *Industry 4.0* and *Industry 5.0* [11] are driving engineering toward the integration of heterogeneous technologies to address the safe coexistence of humans and robots within the same workspace. The revolutionary concept of *cyber-human system* [2] extended to industrial scenarios carries a wide range of opportunities, one among all the realization of tasks with a high degree of flexibility that leverage the capabilities of both the involved agents. In particular, a desirable feature for a collaborative industrial automation cell is the coexistence of human's experience and fine manipulation skills together with a robotic system that is capable of handling large inertia payloads.

The authors want to stress the importance of human operators in collaborative applications and the valuable benefits that experience, dexterity, and fine manipulation skills can add to the task. On the other hand, several tasks related to industrial assembly chains are often repetitive and require physical effort even for long time slots, so optimization efforts are being done for improving ergonomics and reduce fatigue and stress in the operator. Hence, a robotic system can be used to absorb the mechanical effort which affects the human, for instance, completely absorbing the weight of a payload while the operator performs assembly tasks. Of course, continuous collaboration between a human and a heavy-payload industrial robot raises evident safety concern due to the high forces that the system can exert on the environment.

Assembly chain robots are typically isolated inside cages or placed in areas that are restricted to humans to prevent high-speed moving links from having potentially fatal consequences. Bringing industrial robots "*out-of-the-cage*" requires advanced sensing architectures to guarantee the safety of workers and the whole workspace. Cobot architectures are definitely an inspiring model but are also built ad-hoc for physical interaction purposes, while industrial manipulators are designed to achieve high performances in motion accuracy and speed, as well as manipulation of large inertia objects. The chapter presents an architecture that enables collaboration with a manipulator which is not built for human-robot interaction purpose. The application is presented as an example of *robot retrofitting* in the sense that a noncollaborative system can be reconditioned to be collaborative by leveraging a proper sensing architecture.

In practice, the aim of this chapter is to guide the reader through the concept, the design, the implementation, and the testing of an external sensing architecture for performing safe pHRI in *shared payload tasks*, a novel kind of industrial application in which continuous physical contact between robot and operator is strictly required. The described work has been carried out as part of the *H2020 CoLLaboratE* project [12], and the collaborative windshield manipulation is one among the practical scenarios chosen as demonstrators. As a major contribution, this chapter provides a comprehensive description of a technological transfer process aimed to address a practical problem of particular interest for industrial manufacturing. The goal is to provide the reader with a full picture of the application and its motivations and then describe the implementation by following a bottom-up approach, also presenting contents of previous results accomplished by the authors [13–15]. Another goal of this chapter is to describe the concept and implementation of a *shared payload task*, already introduced in References 14 and 15, which represents a novel industrial task for which physical interaction is a necessary requirement. Moreover, the chapter contains a section related to the mechanical realization of the novel gripper structure. The rest of the chapter is structured as follows: In section 7.2, a complete description of the target problem related to the use-case scenario is given. In section 7.3, we present research background related to *pHRI* by reviewing the related methods and technologies. The background review will then cover tactile sensing and tactile data processing which are the main methodologies involved in our research. In section 7.4, a comprehensive description of the developed solution is presented. The design of the tactile device is to be presented from both mechanical and electrical point of view. This section also shows the data-driven methodology adopted to accomplish the voluntary interaction recognition task. Section 7.5 describes the whole gripper architecture developed for the *CoLLaboratE* project use-case from a system perspective, and section 7.6 provides experimental results related to algorithm validation and system testing. Finally, section 7.7 provides final remarks and insights on future work.

7.2 Problem description

The purpose of the research described in the chapter is to design, implement, and test an external sensing architecture in order to adapt an industrial robot arm to perform collaborative assembly tasks. In particular, we are considering as the target problem the use-case of the *H2020 CoLLaboratE* project [12] which consists of a windshield assembly/inspection task, shown in Figure 7.1, where the operator has to handle a large-inertia payload while the robot is absorbing the weightlifting effort.

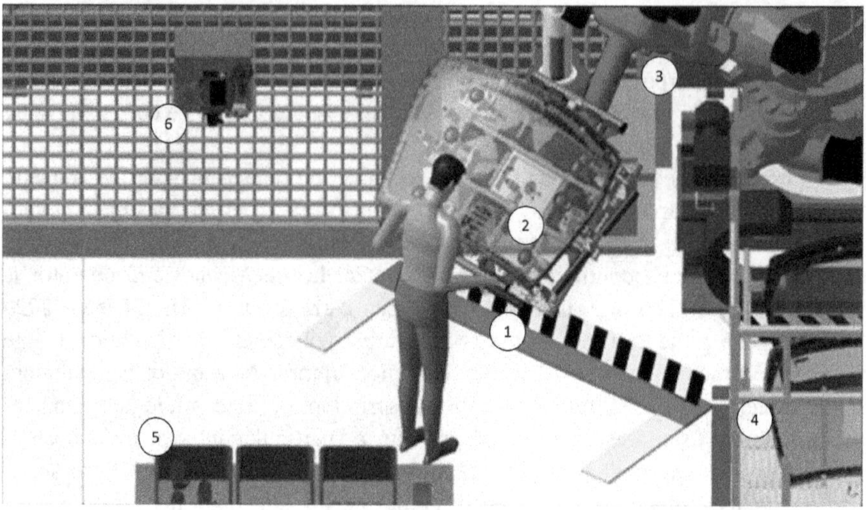

Figure 7.1 *Conceptual visualization of the considered use-case, consisting of a human-robot collaborative windshield handling. The task goal is to detect cracks and fabrication defects, as well as to assemble onboard sensors (i.e., cameras and rain detection) and the rearview mirror. The operator's safety is enforced by the presence of a boundary (here traced on the floor) which represents the limits of the safe zone in which he/she can work. (1) Handles covered with tactile sensors are deployed on three sides of the gripper (frontal, left, and right) to allow the user to grab the gripper from different angles. (2) The windshield is held by a vacuum gripper made of one main sucker located in the center plus a symmetrical set of small suckers holding the corners. (3) Industrial six-dof manipulator. The gripper is rigidly attached to its flange and mounted on top of a force/torque sensor. (4) Racks holding windshields. In the first phase of the task, the robot will collect the part with the vacuum gripper to move it toward the interaction zone. (5) Some bins containing the assembly parts are located close to the operator. (6) Emergency buttons, commands for task phase switching, and teaching pendant.*

This practical problem is clearly an instance of the previously described *shared payload tasks* and represents a fairly complete example of industrial task in which the strengths of both counterparts (machine and human) are merged to enable operations that are not intended to be performed individually by the operator or by the robot.

The considered task necessarily requires the human to engage a physical contact with the robot in order to move the gripper, thus the payload, in the 3D Cartesian workspace. Employing the robot as a weightlifting support requires its motion to be compliant with any action applied to the payload by external agents, providing an adequate system response in terms of damping and overall smoothness. Anyway, the design of a robust motion controller is not enough to guarantee safety in such a critical scenario. This shared payload task requires a deliberative policy that provides a mechanism to smoothly enable the robot motion with respect to specific interaction conditions. For instance, if the operator stumbles and hangs on the payload to avoid falling, an admittance controller would generate fast motions along the fall direction resulting in both the payload and the human hitting the ground, eventually damaging the robot and the floor.

This is also true in case of an accidental contact like an elbow stroke while the worker turns to grab a tool from the table: the robot will be stopped only by the damping provided by the controller but without any warranty on the motion direction of the payload, so other humans or objects can be hit before the control system's response decays. These considerations suggest the need for a method to discriminate when a contact is voluntary or accidental, thus enabling the robot controller only when physical interaction is recognized as intentional by the system. Therefore the sensing architecture must be capable to detect external forces (for motion control) but also to provide a meaningful interpretation of the effective nature of the physical interaction. We have encoded the *voluntary interaction* recognition problem as the detection of at least one human hand touching the gripper because the grasp is the action that better represents a deliberate physical interaction in this context. The extraction of meaningful features which can correctly represent a human hand requires a spatial representation, which cannot be provided by force–torque sensors or vision. In particular, force/torque sensors are widely used in robotic applications due to their reliability and conformity to the most modern standards, but they can only measure forces and torques at a single specific contact point, without providing any insight on the geometry and so the nature of the contact. Vision is also adopted as the main sensory information in human-robot interaction, but it better suits a context of asynchronous cooperation in which no physical interaction is considered. For instance, RGB-D cameras are used to track human motion and monitor the pose of the operator, eventually aborting the motion of the robot when a potential danger is predicted. A comprehensive and dense 3D representation of the scene is desirable in terms of safety, but vision cannot provide any insight about interaction dynamics and contact geometry, and so a standalone vision-based architecture is not suitable for the application described here.

To achieve voluntary interaction detection, we propose the integration of tactile sensing devices on the robot gripper in order to extract geometric data and to provide a representation that encodes both the dynamics and the spatial geometry of the contact interaction. We exploit tactile data to build a spatial representation of the contact, which is then used to discriminate between voluntary and nonvoluntary interaction using data-driven techniques. In particular, tactile sensors allow us to represent contacts over a region as a stream of 2D images which can be treated with well-known image processing and classification methodologies such as *convolutional neural networks* (CNNs) [16]. The output of the whole process is the likelihood that a human hand is correctly recognized in the image, in other words, the probability that the operator is voluntarily grasping the sensing devices.

7.3 Background

The sensing architecture described in this chapter has been designed to enable an industrial robot for physical human interaction with an operator and involves technologies and methods from different fields in engineering and computer science. The following section will present the scientific background of the adopted methodologies which have been leveraged to design the sensing architecture.

7.3.1 Physical human-robot interaction

pHRI is described as a form of *HRI* in which the human and the robotic agent have single or multiple direct body contacts [5]. By its own nature, it yields several problems, mostly related to safety: an undesired contact of a moving heavy robot with a human may easily violate the limits imposed by industrial standards [17]. Some works, e.g., References 18 and 19 address the physical interaction by means of *admittance control* strategies relying only on feedback provided by joint level torque sensors and tested on small-sized robots, but this is unfeasible when dealing with heavy industrial manipulators since joint level sensors are not a common design choice. Moreover, when considering a heavy-duty robot, the detection of a light contact represents a challenge considering the high inertia values of the links.

These considerations highlight the necessity for a *pHRI* interface that can be added to the robot in a modular way and in a position that could be reached by the operator. In particular, an interface between human and robot that does not have strict requirements on its placement could give a valuable benefit in terms of flexibility. As an example, in our application, the sensing devices are rigidly attached to the payload in order to encourage intuitive manipulation, but the same device could be placed virtually everywhere on the robot according to task specifications. Also, the shape of the device is the result of a task-driven design, but as shown in References 13 and 20, distributed sensing technologies allow coverage of heterogeneous and complex shapes without affecting their capabilities.

7.3.2 Devices and systems for pHRI

The literature regarding systems for pHRI presents a broad range of systems that are applied in several areas of robotics.

A safe and comfortable physical interaction is desirable of course in rehabilitation applications, both for the upper [21, 22] and lower limbs [23, 24], as well as in human power augmentation applications. Under the cited domains, the whole robotic system has to be designed ad-hoc for being physically interactive, both from a mechanical and electrical point of view, and should be endowed with built-in sensors to measure interaction forces and any other variable useful for accomplishing the task. Safe human-robot interaction is also broadly studied for industrial applications [5], in which most of the involved robots are not designed and built with a "human-centered" approach and are conceived mostly to perform prefixed tasks in areas that are inaccessible for the operators. This raises the problem of adapting the already deployed robots to human presence through external sensing systems and devices. Most of the solutions shown up to now rely on the usage of cameras and/or motion capture systems like in References 25–27. Vision-based solutions are relatively simple to implement because there is no need to extend the robot structure but suffer from two main limitations. First, vision systems are affected by light changes and occlusions, which can create "blind areas" near the robot or the operator. Second, vision obviously has limited or no capability in measuring physical interaction phenomena which can arise during a collaborative task.

Hence the need for external devices which can endow the robot with physical interaction capabilities. Until recently, few works which address *manipulator retrofitting* for pHRI can be found in the literature [28, 29]. A quite recent solution proposes the design of a compliant wrist [28] based on a spring mechanism and a multisensor architecture comprehensive of contact and proximity sensors. The device has shown good results in enhancing the compliance of a robot, but it is designed to be placed exclusively on the robot's flange, and as remarked by the authors, the proximity sensors lead to low scalability of the system. Moreover, this approach does not consider any kind of higher-level information for differentiating the behavior of the robot during operation. Another more recent approach is presented in Reference 29, where a force/torque sensor is placed under the base of a manipulator and used to estimate contacts all over the robot body.

The estimation techniques rely on data-driven methods and are tested on a custom-built lightweight manipulator. However, the described solution will encounter practical limitations in case of deployment on an industrial robot. The placement of an F/T below its base would be impractical, and estimation algorithms cannot provide an exact localization of the contact points given the base torques, especially in the presence of heavy links. To endow industrial manipulators with physically interactive behaviors, purely reactive compliance-based approaches could not be enough to insure safety standards [3]; therefore, an interface also capable of providing a higher-level abstraction of interaction data should be desirable. A recent paper [30] presents a device similar to the one considered in this chapter. A cylindrical handle covered with distributed tactile sensors is used for the semantic interpretation of basic grasp configurations. Thus, no discrimination between voluntary and nonvoluntary interaction is made. The computational approach is based on images processed with CNNs and aims at a classification that corresponds to different motions of the manipulator. The experiments are performed with a Cobot, namely a UR3 model, and the handle is attached directly to the flange. The authors aimed mostly to provide a semantic interpretation of the interaction for generating open-loop motion commands, but they suppose that physical contact is always performed by a human. Indeed, previous considerations highlight the need for high-level interpretations of physical interaction data achievable using distributed tactile sensors.

7.3.3 Tactile sensing in pHRI

As remarked before, when treating *pHRI* problems, it is desirable to have a semantic interpretation of interaction data. A natural way to achieve it is through the sense of touch: it provides richer physical information compared to force/torque sensors and joint level sensors embodied in a manipulator. Moreover, tactile sensors can be distributed over flat and curved surfaces leading to multiple simultaneous contact sensing. In literature, several tactile sensing systems have been presented based on different transduction technologies: capacitive [31, 32], piezoelectric [33, 34], and optical [35, 36]. Most of them aim at the reconstruction of a pressure distribution, although some works go even further proposing multimodal systems capable to provide also acceleration and temperature data [37]. Tactile sensors measurements have been used in the context of HRI and pHRI in order to implement touch-based control strategies. In Reference 38, the authors show that patches of pressure sensors integrated on a robot link provide a manual guiding interface that leverages pressure readings to define motion vectors used for controlling the arm position. Similarly, Schmidt *et al.*, in Reference 39, used an array of capacitance-based pressure sensors mounted on a robot gripper in order to drive the robot posture in response to the force applied by a human operator. The work described in Reference 40 shows the implementation of a compliant behavior for a robot arm exploiting the feedback of a force-sensitive array. Whole-robot body compliance was achieved in Reference 41 by using a technique based on hierarchical force propagation exploiting force feedback provided by an artificial skin. The work in Reference 20 proposes a touch-triggered task-based control method using robot skin tactile feedback allowing a human operator to physically drive robot motions in Cartesian or joint space.

From a semantic point of view, tactile sensors have been also used to recognize different touch modalities, namely actions (e.g., pat, push, etc.) performed by human subjects using the hand. The general approach is similar to most of the techniques proposed in the literature: a set of features is extracted and classified using supervised machine learning algorithms. The main differences among the various solutions are related to the number of modalities classified and the training methodologies adopted. In particular, in Reference 42, the authors used a k-nearest neighbors algorithm to classify five touch modalities based on data collected in experiments involving 11 users. A neural network has been considered in Reference 43 in order to classify a set of eight interactions performed by a single subject. The work described in Reference 44 shows the implementation of a support vector machine to recognize nine touch modalities using a multimodal robot skin providing pressure, acceleration, and proximity measurements. In all the works discussed above, it is implicitly assumed that a person is interacting with the robot: namely, all the contacts used for the classification have been generated by humans. Therefore, they all have not been addressing the possibility of discriminating human touch from other possible types of contacts. Recent works described in References 13 and 45 address the possibility of discriminating human touch from contact events of different nature. In particular, distributed contact sensing is used to distinguish between voluntary and involuntary physical contact using salient features by analyzing the shape of the contact pressure distribution. A review of the methods and techniques for contact shape processing is presented in the following.

7.3.4 Contact shape classification

The application presented throughout this chapter requires the processing of spatial tactile information to extract information about the nature of the contact. In applications requiring the processing and classification of the contact shape, it is common to convert the pressure data distribution into a tactile image, a representation that maps the intensity of each pixel to a pressure value. By following this rationale, tactile images can be processed by leveraging well-established image understanding algorithms. As an example, the authors of Reference 46 show that tactile images generated from a fingertip can be used to classify in real-time primitive shapes and poses of the contact. In Reference 47, the authors covered a robot hand with small planar tactile patches mapping the whole pressure readings onto a single image. Finally, they trained a neural network to classify a set of grasped objects. Cao *et al.* in Reference 48 used a stream of tactile images obtained during a grasping task to classify ten different objects using a CNN. The work described in Reference 49 proposes an approach where a high-resolution patch of pressure sensors integrated on a gripper is used to classify the tactile images generated by objects, human limbs, and fingers through a CNN. All the mentioned works prove that shaping tactile data as images yield effective advantages for their processing and interpretation. In our work, we leverage the methodologies and models described in Reference 20.`

7.4 A novel tactile device for pHRI

The presented work aims to provide safe physical interaction capabilities to an industrial robot built for holding heavy payloads and performing fast and precise motions in space. The robotic system considered here is designed without taking into account any kind of interaction task, physical interaction above all. Therefore, to safely make contact with such a robot requires a proper external sensing architecture to interface the manipulator with the external environment, and in particular with the human working with it. We imbed distributed tactile sensors on the gripper through ad-hoc physical interfaces, which we have represented as cylindrical handles designed to be comfortably grasped by the operator. This section describes the design and implementation of the sensing handles used to recognize voluntary physical interaction.

7.4.1 Mechanical design

The cylindrical layout of the device has been chosen because it represents a natural shape to grasp thinking about common objects (i.e., doors and cups). The mechanical structure of the device is represented in Figure 7.2 and has the following salient features, highlighted by their numeric label in the figure: Its external cylindrical structure is composed of two identical halves (1) made of polylactic acid and realized with 3D printing technology. This design choice aims toward easier maintenance operations and an overall simplicity of electronic integration since the sensors are wrapped around the two halves. The internal shape of the cylindrical halves has narrow empty sections, (2) needed for the electrical integration of the sensors since this operation requires signal and power cable routing as well as the placement of dedicated communication devices, which will be presented in section 7.4.2. The core mechanical structure of the device is a carbon cylinder (3) which provides

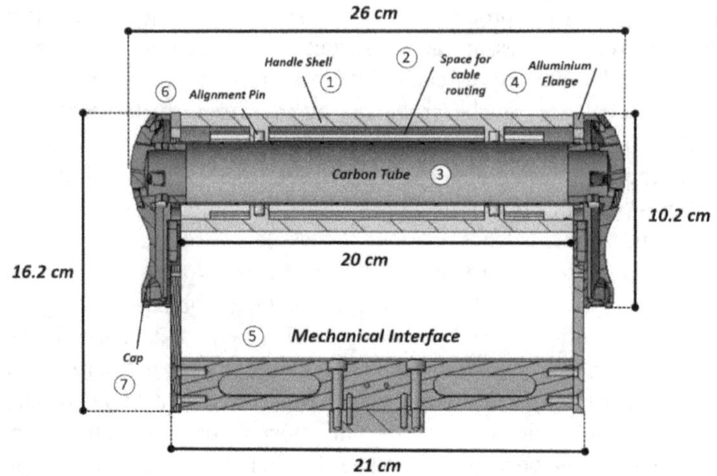

Figure 7.2 Longitudinal section of the CAD model of the device [14]

mechanical support for the printed handles. It exceeds by 25 mm at each side with respect to the shell length of 200 mm so that it can be lodged in a slot bored in the aluminum flanges (4). The flanges are made of aluminum and realized with a milling machine; they are used to connect the inner tube with the aluminum mechanical interface (5) which allows the integration of the gripper. More details on the mechanical structure of the gripper are reported in section 7.5.

The two halves of the shell are fixed to the flanges through screws for preventing undesired opening of the handle, and rotation about the tube axis is prevented by using cylindrical steel pins (6) which pass through holes bored on the tube surface. Figure 7.2 also shows two 3D-printed structures (namely *caps*) that are used to seal the cylindrical structure on its sides (7). The caps also can host a small amount of tactile sensors, even if the available surface is not enough to detect hand-sized shapes. Nevertheless, they can provide contact localization and force information along the cylinder main axis, which is not exploited for this particular application, but it represents a useful feature for tasks in which contact forces are required. Having sensors on the handle halves allows to measure only the radial forces since the adopted sensors can provide measurements only along the normal vector to their surface, so the integration of tactile modules also on the caps increases the span of measurement directions.

Our choices related to the sizing of the device are motivated by the following assumptions:

- The cylindrical shape of the handle has a length of 200 mm. It has been chosen since it is simple and easy to grasp: in fact, it insures a natural grasping pose for the human hand.
- The external diameter of the handle is 64 mm as result of a trade-off; this size is larger with respect to the optimal diameter expected for a cylindrical tool [50]

but allows a better placement of flexible sensors and internal cables, anyway ensuring a firm grasp.

- The internal tube has a length of 205 mm and a diameter of 34 mm and is used as support for the sensing handle when attaching it to the mechanical interface of the gripper. It slightly exceeds the length of the outer cylinder because it fits in two circular slots located on the surface of the flanges, ensuring better stability during the mounting phase.

7.4.2 Electrical design

The two halves of the handle are covered with tactile sensors based on CySkin technology, presented in Reference 32. It consists of an artificial skin composed of distributed tactile sensing elements (i.e., *taxels*) measuring the pressure distribution generated by the physical interaction. The skin is composed of flexible interconnected triangular modules that can be cut and adapted to curved surfaces. Each module can contain up to 11 capacitive-based pressure transducers. The pitch among nearby taxels is 8 mm. In the proposed application, we integrated 320 taxels in each half of the handle, which insure coverage of the cylindrical surface (see Figure 7.3). Such coverage allows to capture the whole operator hand while interacting with the gripper and therefore to reconstruct its contact shape.

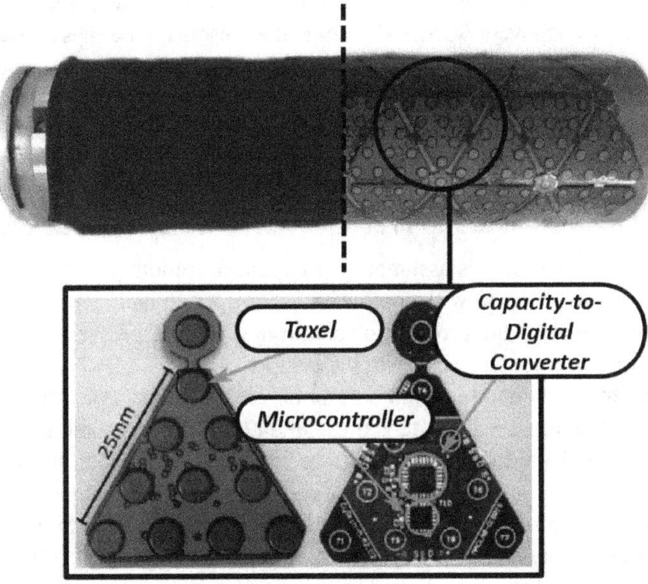

Figure 7.3 CySkin array placed on the upper half of the handle, with and without the lycra layer. The figure shows the detail of a single triangular module composed of 11 taxels. On the back side are integrated the capacity-to-digital converter and the microcontroller which drives the module [14].

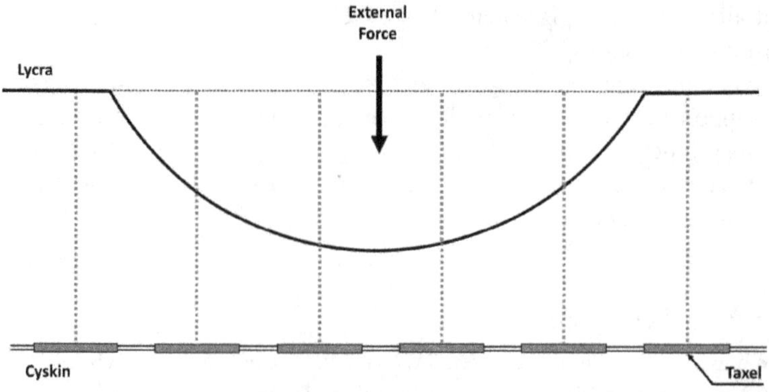

Figure 7.4 *Schematic model of the capacitive transduction principle of CySkin [32] based on the deformation of the conductive lycra layer*

CySkin working principle is based on capacity variation, measured by dedicated circuitry integrated on the back of each triangular module. A *capacity-to-digital converter* is used to detect variations in the capacity between the circuit layer on which the taxels lie and a layer of conductive lycra used as a ground plane, wrapped around the two halves and fixed by Velcro strips placed in the internal space. Measured variations are induced by the deformation of the lycra layer as shown in Figure 7.4 and allow to retrieve a pressure distribution. Two small lycra patches also cover the sensors on the caps and are fixed by two printed masks. An integrated processing unit (**I**ntelligent **H**ub **B**oard [32, 51]) placed inside the handle collects the data from the skin patches through *SPI (serial periphera interface)* protocol. Data are then sent to a host PC via *CAN BUS* protocol, providing measurements at a frequency of 20 Hz.

7.4.3 Voluntary interaction recognition through hand detection

In the use-case scenario, it is assumed that to start a voluntary interaction with the robot, the human operator will approach the gripper grasping it with its hands. Therefore, we consider the problem of classifying the pressure distribution generated by a human hand interacting with the gripper.

We have modeled hand detection as an object detection problem, so we represent tactile data as images. The steps presented here briefly resume the image generation process and are described in depth in the works cited along the subsection.

A 3D mesh is computed by exploiting the set of taxels positions and the *Delaunay triangulation algorithm*. The mesh is exploited to obtain a 2D tactile image that best preserves all the spatial relationships found in its 3D counterpart. The flattening algorithms exploit the results of the *surface parameterization theory* and are described in References 13 and 52. When running data acquisition, at each measurement step, a snapshot of the raw pressure values of each taxel is sampled; thus, a grayscale tactile image is computed by following the pipeline described in Figure 7.5, where the grayscale intensity levels of the pixels are modulated by the perceived forces.

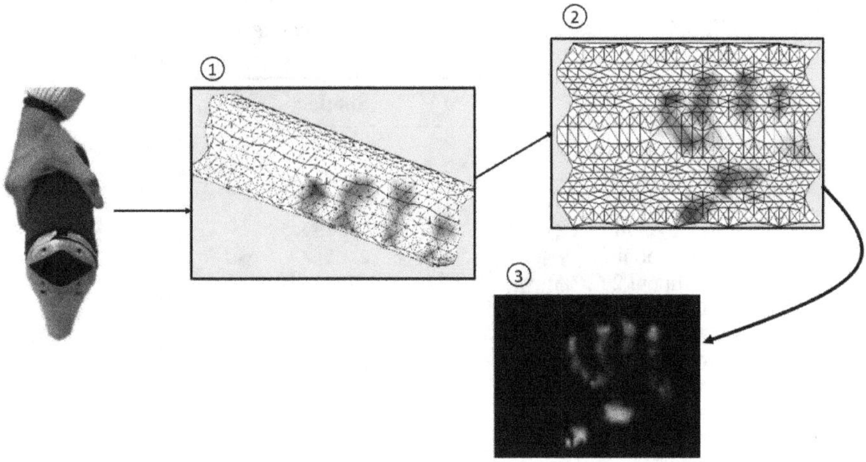

Figure 7.5 *Tactile image generation pipeline. The figure shows how a grasp action is processed to be encoded as a tactile image. The 3D pressure distribution represented on the mesh (1) is flattened exploiting an algorithm described in References 13 and 52 to obtain a 2D tactile map (2). The map is then interpolated to generate the tactile image (3)[14].*

In our application, human hand detection is modeled as a binary classification task that can be tackled by using machine learning techniques suitable for image classification tasks, such as CNNs. The hand detection problem has been tackled by our research group in Reference 52 and then in Reference 13, works which describe the implementation of *HandsNet*, a CNN trained for the detection of human hands from tactile images. The HandsNet architecture used in this work is the one reported in Reference 13: the backbone of the network is composed of four convolutional blocks, each one containing a convolutional layer, a padding layer, and a batch normalization layer [53], ending with a rectified linear unit activation function [54]. The output of each layer is downsampled by using MaxPool filters. The complete network structure is resumed in Table 7.1. In this work, we have considered an instance of HandsNet, which was pretrained by acquiring tactile images on a Baxter Robot link. Toward the shared payload handling scenario, we performed a fine-tuning to adapt the *HandsNet* model trained on the dataset presented in Reference 13, acquired on a sensorized robot link, to work on the novel sensing devices. This step is required since the process of converting the tactile measurements into images introduces distortions that are dependent on the geometry of the surface where the skin is integrated [13]. Moreover, the model must be also capable of recognizing when a contact is not associated with a grasp action in order to stop robot motion. We have identified undesired physical interaction both with excessively large areas or very small contact areas. The first could represent a situation in which the operator is colliding with an arm/forearm. The second situation represents a contact made with only few fingers, which could mean transient situation in which the handle is being grasped or released.

Table 7.1 HandsNet structure as described in Reference 13

Layer	Shape
Convolutional 1	32 × 7 × 7
Max pool 1	2 × 2
Dropout 1 (10%)	—
Convolutional 2	64 × 5 × 5
Max pool 2	2 × 2
Dropout 2 (20%)	—
Convolutional 3	128 × 3 × 3
Max pool 3	2 × 2
Convolutional 4	256 × 3 × 3
Max pool 4	2 × 2
Fully connected 1	64
Dropout (60%)	—
Fully connected 2	32
Dropout (50%)	—
Fully connected 3	2
Softmax	2

7.5 The CoLLaboratE gripper architecture

The target use-case for the tactile-based gripper architecture is a car windshield assembly and visual inspection in a human-robot collaborative scenario. On this purpose, ad-hoc mechanical and computational architectures have been integrated on a vacuum gripper, a system commonly adopted in car manufacturing. Typically a vacuum gripper exploits a pneumatic actuation mechanism to exert suction force on the payload allowing to hold it firmly without damaging or scratching the surface. Here, the vacuum architecture is also exploited to actuate a compliant mechanism built to deploy the tactile handles during operation. The CAD models presented in Figure 7.6 give an overview of how the gripper and the handle deployment mechanism are structured.

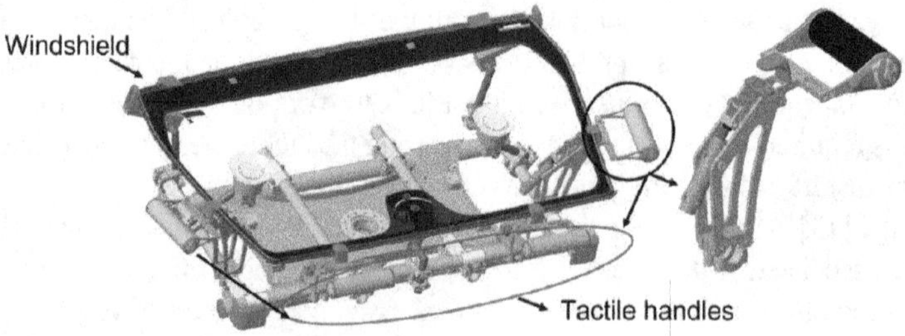

Figure 7.6 CAD model of the gripper with mounted handles. The pneumatic-actuated handle support is shown on the right [14].

The need for an actuated support arises from the structure of the task. Up to now, the chapter focused more on the collaborative aspect, which is fundamental for the inspection and refinement of the windshield. Anyway, before the interactive phase, the large payload must be retrieved by its static support and moved near the operator, which stands in a specific zone of the workcell. Due to safety reasons, the worker cannot move freely inside the cell but waits for the payload to be ready on the robot's flange in a safe zone located within the reach space of the robot but far from its base enough to avoid collision with nonsensorized links. During the first phase of the task, the robot moves autonomously to grab the windshield and to bring the gripper close to the operator in the interaction zone. Thus, the handles are not required, and a potential collision with an unmodeled obstacle even just on the edge of the gripper could lead to damage the sensors. The pneumatic mechanism is used to keep the handles in a safe configuration so that they cannot be harmed by collisions during the payload retrieval operations and in general in any noncollaborative section of the task. The handles will be deployed only during the collaborative phase of the task where the human starts inspecting the windshield for visual cracks and starts assembling rear-view mirror and sensor components on the windshield. Four handles are deployed simultaneously to cover the boundary and provide the human operator any easy access to the tactile sensors. The handles enable the operator to manipulate the robot in Cartesian space during the human-robot collaboration phase of the task. Apart from hand-guiding operation, higher-level abstract tasks to indicate the start of pHRI, completion of the collaborative phase can be easily communicated to the main robot controller without leaving the safe zone of the workcell through the tactile handles. In Figure 7.7 a comprehensive scheme of the system is provided.

7.5.1 Computational architecture

The gripper imbeds computational units for acquiring and processing data from both tactile and force/torque sensors. The pressure measurements from the tactile sensors travel over CAN BUS protocol, thus a multi-input CAN hub is used to collect raw data from the handles. The CAN hub performs a conversion to USB which is sent to a computational unit for further processing. We chose to use a *NVIDIA Jetson Nano,* which is a small general-purpose board with GPU unit. The presence of the GPU allows to directly compute predictions by running the *HandsNet* model on it. The presence of a small computational unity enhances the flexibility of the gripper architecture, which in fact represents a complete system providing the sensors, the early data processing and also software modules for semantic interpretation of the processed information.

Also, the data provided by the force/torque are collected by the *Jetson* via Ethernet. Moreover, the board is used to send basic commands to the pneumatic actuation system, which drives the behavior of the suction cup and of the handle deployment mechanisms.

7.5.2 Mechanical design

A specialized gripper was designed and developed to encapsulate all the features required for a collaborative robotic assembly. The model was conceptualized based on widely used automotive windshield gripper design, where vacuum suction is

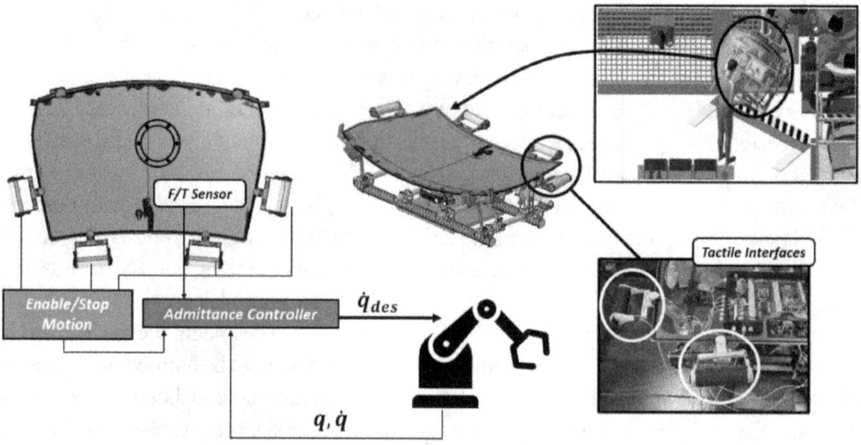

Figure 7.7 Diagram of the gripper architecture for the use-case scenario. The robot control scheme is based on an admittance law whose behavior is enabled or disabled whether voluntary contact with handles is detected. The admittance controller leverages measurements both from robot's internal sensing and from the force/torque sensor mounted on the gripper. q is the array of joint positions, and $\dot{q}$ is the array of joint velocities. $\dot{q}_{des}$ represents the desired joint velocities at each time step.

used to grab and hold the part. A handle-like structure is proposed and designed as shown in Figures 7.6 and 7.8. The handles are intended to be available only during the human-robot collaboration phase, where the human operator starts visual inspection and assembly, as they are protruding structures that cannot be suitably encompassed within the boundary of either the windshield or the gripper. Additionally, there could be interference with the surrounding during the grasping phase and collision with the chassis during final assembly phase. Hence, the handles are mounted on a deployable structure which is made available only during the interaction phase. The major characteristics for the functional requirement of the HRI handle depend on the two extreme positions: when the mechanism is in open position (deployed) and when in closed position (retracted). When the mechanism is in open position, it is desirable to follow specific design rules related to the handles:

1. They should be in a comfortable position for the operator.
2. They must not be an obstacle to other assembly activities.
3. As a safety feature, in case of power failures, the mechanism should be in the open position.

Several deployable mechanisms are described in the literature such as consist of scissor-like elements [55], Hoeken linkage [56], Sarrus linkage [57], and metamorphic mechanisms [58]. A six-bar planar mechanism similar to Watt-II linkage [59]

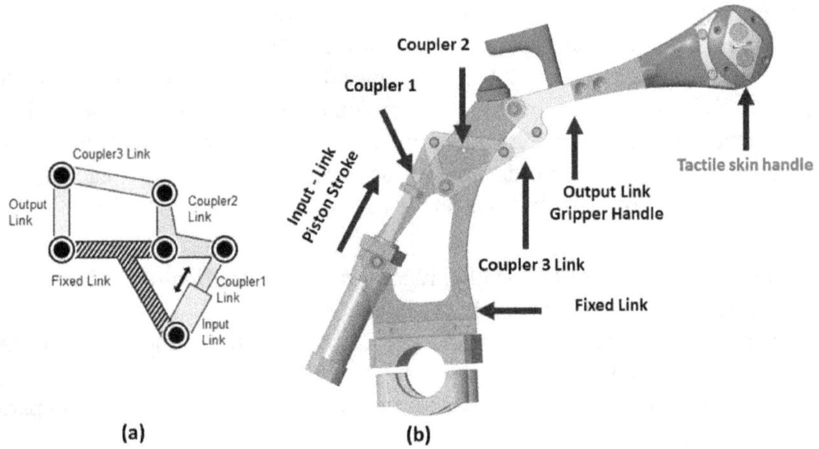

Figure 7.8 *Deployable mechanism based on six-bar planar linkage is shown. (a) Kinematic linkage. (b) Tactile handle unit is represented where the (i) input-link is mounted on pneumatic piston, (ii) coupler 1 connects vacuum piston and the coupler 2, (iii) coupler 2 is connected to the fixed link mounted to the gripper, coupler 1 link and coupler 3 link, (iv) coupler 3 link is connected to the output link, and (v) tactile handle is mounted on the output link.*

was designed to implement the deployable system. The mechanism has two-closed loops, where one loop consists of pneumatic piston actuation and the other consists of the tactile handles. The fixed link for the linkage system refers to the rigid structure attached to the gripper frames as shown in Figure 7.8.

In addition to handles deployment, the mechanism also provides sufficient damping to the force exerted by the operator on the tactile devices. The pneumatic pistons in this case enable the tactile handles to withstand forces up to 200 N.

7.6 Bench testing and experimental results

The sensing devices have been bench-tested before their deployment on the gripper. After the preliminary tests related to their proper operation, one of the four handles was employed to acquire a new dataset for the fine-tuning of the *HandsNet* model. Experiments related to the performances were performed to evaluate the inference time of the onboard computational architecture to assess the reliability of the solution.

7.6.1 Data acquisition

Data acquisition sessions have been performed to collect a new dataset of tactile images, generated only by physical interaction with the device.

As previously mentioned, fine-tuning is required to adapt the *HandsNet* model described in Reference 13 on the handle geometry. The new set of tactile images represents physical contacts involving the sensors placed on the handle. The data collection involved the participation of 18 subjects. The group consisted of males and females of the age between 22 and 30. Since the pressure distribution generated by the human hand depends (in terms of magnitude and shape) on the type of physical interaction performed by the human operator, we asked the subjects to perform different grasping actions in order to consider this aspect. Therefore, each subject had to perform the following actions on the handle:

- ten standard grasps of the handle, regardless the orientation of the hand (they were encouraged to alternatively use both hands)
- two horizontal pushes of the handle, perpendicular to the cylinder axis and parallel to the ground
- two horizontal pulls of the handle, perpendicular to the cylinder axis and parallel to the ground
- one left pull, parallel to the cylinder axis
- one right pull, parallel to the cylinder axis

The chosen actions resemble those which are more likely to happen in the real use-case while the operator is handling the gripper with the windshield, and we took into account both left and right hands because we assume that the interaction is always performed with two hands, respectively, on two different handles. The subject was also asked to perform eight random physical interaction using other parts of the body (e.g., elbow, arm, and shoulder) to simulate collision or more in general nonvoluntary interactions. We also considered grasp action transients (i.e., the time frames in which the user is beginning to grasp or releasing the handle) as *nonhand* examples since in such a safety-critical application, we assume that only a firm grasp can be considered as a completely voluntary interaction. From the collection of tactile images acquired for each subject, we manually chose significant training examples to obtain a balanced set of training examples. We did not consider a fixed time limit for each subject's experiment since for any discrete action, there were differences in each person's approach to the gesture, like the length of the grasp transient or the time period between one grasp and the next one. At the end of the collection procedure, we obtained a dataset composed of 324 *hand* images and 324 *nonhand* images (Figure 7.9), with a resolution of 100×68 pixels and grayscale encoding.

7.6.2 HandsNet fine-tuning

The fine-tuning procedure was performed on a desktop PC equipped with a NVIDIA GeForce RTX 2060 GPU and an Intel i7 processor, which runs Ubuntu 20.04. The dataset of tactile images was split by following a 70%/10%/20% partition for train, validation, and test sets, respectively. The model was retrained for 80 epochs using the *Stochastic Gradient Descent* algorithm, a batch size of 64, and a fixed learning rate of 0.005. As mentioned in section 7.4.3, *HandsNet* is composed of four convolutional layers, two fully connected layers and a two outputs classification layer. During the fine-tuning, the first two convolutional layers have been *frozen* (i.e., they have not been retrained). The third and fourth convolutional layers have

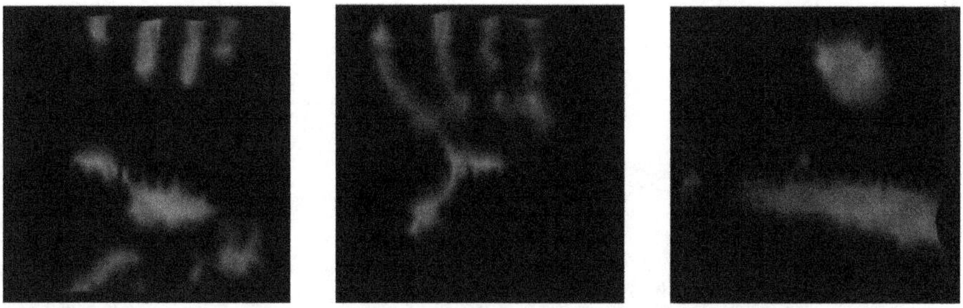

Figure 7.9 *Examples of tactile images. Left: the hand shape is split since the contact occurred over the cut needed to obtain a planar representation of the handle geometry, a situation which was not taken into account in the dataset acquired on the robot link since due to its larger diameter, a human hand could not wrap more than its half. Center: the hand shape clearly visible even if slightly distorted by procedure mapping the handle nonplanar geometry into an image. Right: An example of a contact performed with the forearm on the handle [14].*

been retrained using a fraction (0.25 and 0.5, respectively) of the fixed learning rate so that the propagation of each new training example would have a lower impact on the weights update. The two fully connected layers have been fully retrained. The fine-tuning procedure led to accuracy results of 98.08% and 94.61% on the validation and test sets, respectively. The amount of false positives obtained for the *nonhand* class exceeds the false negatives due to the wider range of possible *nonhand* situations, while the *hand* class examples have a more common and well-defined contact shape.

7.6.3 Inference time performance evaluation

The following section shows a performance comparison between two devices: a desktop workstation with a *RTX 3070 GPU* and a *Jetson Nano* board, both manufactured by Nvidia. Assuming that data are acquired by devices that are placed on the robot gripper, the adoption of a workstation for computing inference requires a more complex communication architecture since the size and weight of the machine do not allow direct placement on the robot. A local network should be used to send data from the onboard sensors to the computing device and then the prediction back to the robot controller, eventually introducing a source of latency. On the other hand, having a small GPU board placed directly on the gripper allows to perform data acquisition and model inference on the same device, leaving only one communication channel toward the controller. We have performed an inference time analysis by running the model on real data acquired by CySkin in order to assess the feasibility of both solutions. Since the presented analysis is independent of the robot motion, we have used a single sensorized handle fixed on a table. Since this study presents an analysis purely focused on detection performances, data have been acquired from a single subject interacting with the device, and a set of

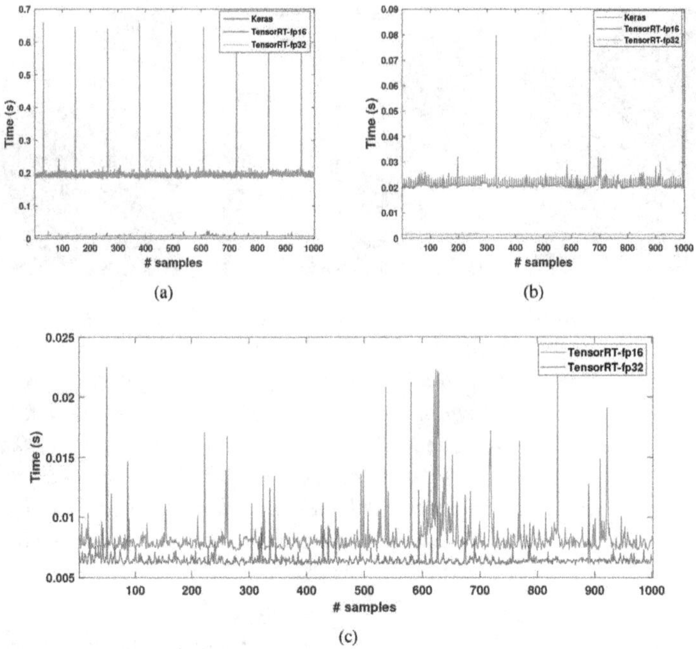

Figure 7.10 *Inference time plots on 1000 samples. (a) The inference time obtained on Jetson Nano. (b) The inference time obtained on RTX 3070. (c) An insight on fp16 and fp32 obtained on Jetson Nano experiments [15].*

1,000 frames has been considered for each experiment. An in-depth user study that also involves robot motion performances is currently being performed.

We also have considered three different instances of HandsNet for each architecture as a complement for our analysis. We have collected inference time from a plain Keras model directly obtained by the framework used for building the network and from two models obtained by compilation with TensorRT (TRT), a toolchain that provides optimization for GPU architectures. The difference between the two optimized models is the floating-point precision used for representing the network parameters: here, we discuss the behavior of *floating-point 16* and *floating-point 32*.

Figure 7.10(a) and (b) and Table 7.2 show that the Keras model exhibits poor time performance compared to TRT models on both devices. Only on the workstation, the time constraint is met by observing the mean inference time. Anyway, it must be noticed that peaks reaching $0.0808s$ are detected, so there is no guarantee that inference can be always performed within the same acquisition step. On both architectures, TRT models have obtained acceptable results in terms of inference time. On RTX 3070, the difference in performance between fp16 and fp32 is negligible, instead on the Jetson board the fp32 model has a mean inference time improvement of $1.8ms$. The fp32 model on Jetson also shows lower standard deviation with respect to the fp16 counterpart. The better quantitative performance and stability of the fp32 model are also shown in Figure 7.10(c).

Table 7.2 Results of the inference time analysis, expressed in seconds (s) [15]

	Keras			TensorRT (fp16)			TensorRT (fp32)		
	Mean	**Max**	**Std**	**Mean**	**Max**	**Std**	**Mean**	**Max**	**Std**
RTX 3070	0.0216	0.0808	0.0038	0.0016	0.0024	9.8×10^{-5}	0.0016	0.0022	1.13×10^{-4}
Jetson Nano	0.1991	0.6786	0.0438	0.0083	0.0224	0.0017	0.0065	0.0225	8.01×10^{-4}

7.6.4 Inference accuracy robustness

The proposed use-case scenario is set in an industrial environment and is focused on physical interaction between the robot and a specialized worker. Voluntary interaction detection is performed by classifying a human hand shape from a tactile image when the human is grasping the devices described in section 7.5. In assembly departments, there are strict safety policies that the workers must follow, especially regarding protective clothing. In the described scenario, the operator is expected to grasp the handles while wearing protective gloves, like shown in Figure 7.11. Since the neural network was originally trained with bare hand grasps, we provide an assessment of its classification performance when grasp is performed while wearing gloves. We acquired data from a user grasping one fixed handle and holding the grasp for 2 minutes while randomly rotating the wrist to emulate the small motions which occur while the robot moves. We analyzed the probability value of the hand class in terms of mean, minimum value, and standard deviation. As shown in Table 7.3, wearing a work glove has a small impact on the prediction value. In particular, the presence of the glove slightly increases fluctuations as can be seen by observing the standard deviation and the minimum value, but in terms of classifications, these values are negligible even by choosing a tight threshold of 0.9 to enable the controller.

Figure 7.11 Detail view of the use-case scenario which shows the two front handles and the work gloves of the operator. The glove model is the same used for robustness experiment [15].

Table 7.3 Results of comparison between hand class probability in bare hand grasp and work glove grasp. Reported results represent probability values; thus, they are dimensionless and limited between 0 and 1 [15].

	Mean	Min	Std
Bare hand	1.00	0.99	5.14×10^{-6}
Work glove	0.99	0.95	0.0019

7.7 Conclusion

In this chapter, we presented a tactile sensing architecture based on grasp interfaces for pHRI. We showed how the described solution can be used to realize a shared payload task in an industrial setting. In particular, by exploiting data-driven techniques, the robot control system can detect voluntary physical interaction. On this purpose, we performed the fine-tuning of the architecture presented in Reference 13 to discriminate between voluntary and involuntary interactions using distributed tactile sensing measurements.

Finally, the classification architecture was implemented on the *NVIDIA Jetson Nano*, a small-sized host PC, that can be easily integrated on the robot gripper. As previously mentioned, the architecture is expected to be tested on the use-case scenario for the *CoLLaboratE* project. Currently, we are pursuing further research related to the sensing architecture. We are planning to use force data from the handles to also provide motion commands to the robot together with the force/torque sensor. Moreover, the development of learning models to interpret the semantics of complex hand gestures on the handles is being investigated.

Acknowledgment

This work was supported by the European Community's Framework Program Horizon 2020 (grant agreement number 820767, project CoLLaboratE).

References

[1] Parodi F., Gerio G.G. 'Aura: an example of collaborative robot for automotive and general industry applications'. *27th International Conference on Flexible Automation and Intelligent Manufacturing, FAIM2017*; Modena, Italy, 27–30 June 2017. Available from https://www.sciencedirect.com/science/article/pii/S2351978917303207

[2] Sherwani F., Asad M.M., Ibrahim B.S.K.K. 'Collaborative robots and industrial revolution 4.0 (IR 4.0)'. *2020 International Conference on Emerging Trends in Smart Technologies (ICETST)*; Karachi, Pakistan, 2020. pp. 1–5.

[3] Bragança S., Costa E., Castellucci I., Arezes P.M. 'A brief overview of the use of collaborative robots in industry 4.0: human role and safety' in Arezes P.M., Baptista J.S., Barroso M.P. (eds.). *Occupational and environmental safety and health*. Cham: Springer International Publishing; 2019. pp. 641–50.

[4] Cherubini A., Passama R., Crosnier A., Lasnier A., Fraisse P. 'Collaborative manufacturing with physical human-robot interaction'. *Robotics and Computer-Integrated Manufacturing*. 2016, vol. 40, pp. 1–13. Available from https://www.sciencedirect.com/science/article/pii/S0736584515301769

[5] Haddadin S., Croft E. 'Physical human-robot interaction' in Siciliano B., Khatib O. (eds.). *Springer Handbook of Robotics*. Cham: Springer International Publishing; 2016. pp. 1835–74.

[6] Universal Robots. Available from https://www.universal-robots.com/

[7] Kuka iiwa Cobot. Available from https://www.kuka.com/en/products/robotics-systems/industrial-robots/lbr-iiwa

[8] El Zaatari S., Marei M., Li W., Usman Z. 'Cobot programming for collaborative industrial tasks: an overview'. *Robotics and Autonomous Systems*. 2019, vol. 116, pp. 162–80. Available from https://www.sciencedirect.com/science/article/pii/S092188901830602X

[9] Pratt G.A., Williamson M.M. 'Series elastic actuators'. *Proceedings 1995 IEEE/RSJ International Conference on Intelligent Robots and Systems. Human Robot Interaction and Cooperative Robots*; Pittsburgh, PA, USA, 1995. pp. 399–406.

[10] Wolf S., Grioli G., Eiberger O., *et al.* 'Variable stiffness actuators: review on design and components'. *IEEE/ASME Transactions on Mechatronics*. 2016, vol. 21(5), pp. 2418–30.

[11] Industry 5.0. Available from https://ec.europa.eu/info/research-and-innovation/research-area/industrial-research-and-innovation/industry-50_en

[12] CoLLaboratE project. Available from https://collaborate-project.eu/

[13] Albini A., Cannata G. 'Pressure distribution classification and segmentation of human hands in contact with the robot body'. *The International Journal of Robotics Research*. 2020, vol. 39(6), pp. 668–87. Available from https://doi.org/10.1177/0278364920907688

[14] Grella F., Baldini G., Canale R, *et al.* 'A tactile sensor-based architecture for collaborative assembly tasks with heavy-duty robots'. *20th International Conference on Advanced Robotics (ICAR)*; Ljubljana, Slovenia, 2021. pp. 1030–35.

[15] Grella F., Canale R., Giovinazzo F., Albini A., Cannata G. 'Tactilebased human-robot collaboration: a performance analysis' in Valle M., Lehmhus D., Gianoglio C. (eds.). *Advances in system-integrated intelligence*. Cham: Springer International Publishing; 2023. pp. 437–45.

[16] Rawat W., Wang Z. 'Deep convolutional neural networks for image classification: a comprehensive review'. *Neural Computation*. 2017, vol. 29(9), pp. 2352–449.

[17] Bicchi A., Peshkin M.A., Colgate J.E. 'Safety for physical human-robot interaction' in Siciliano B., Khatib O. (eds.). *Springer Handbook of Robotics*. Berlin, Heidelberg: Springer; 2008. pp. 1335–48.

[18] Magrini E., Flacco F., De Luca A. 'Control of generalized contact motion and force in physical human-robot interaction'. *2015 IEEE International Conference on Robotics and Automation (ICRA)*; Seattle, WA, USA, 2015. pp. 2298–304.
[19] Keemink A.Q., van der Kooij H., Stienen A.H. 'Admittance control for physical human-robot interaction'. *The International Journal of Robotics Research*. 2018, vol. 37(11), pp. 1421–44. Available from https://doi.org/10.1177/0278364918768950
[20] Albini A., Denei S., Cannata G. 'Enabling natural human-robot physical interaction using a robotic skin feedback and a prioritized tasks robot control architecture'. *2017 IEEE-RAS 17th International Conference on Humanoid Robotics (Humanoids)*; Birmingham, IEEE, 2017. pp. 99–106.
[21] Mihelj M., Nef T., Riener R. 'ARMin – toward a six dof upper limb rehabilitation robot'. *The First IEEE/RAS-EMBS International Conference on Biomedical Robotics and Biomechatronics, 2006. BioRob 2006*; Pisa, Italy, 2006. pp. 1154–59.
[22] Zimmermann Y., Forino A., Riener R., Hutter M. 'ANYexo: a versatile and dynamic upper-limb rehabilitation robot'. *IEEE Robotics and Automation Letters*. 2019, vol. 4(4), pp. 3649–56.
[23] Zhang F., Hou Z.G., Cheng L., *et al.* 'A lower limb rehabilitation robot: a proof of concept'. *IEEE Transactions on Human-Machine Systems*. 2016, vol. 46(5), pp. 761–68.
[24] Saglia J.A., Luca A.D., Squeri V., *et al.* 'Design and development of a novel core, balance and lower limb rehabilitation robot: hunova®'. *2019 IEEE 16th International Conference on Rehabilitation Robotics (ICORR)*; Toronto, ON, Canada, 2019. pp. 417–22.
[25] Tellaeche A., Maurtua I., Ibarguren A. 'Human robot interaction in industrial robotics. examples from research centers to industry'. *2015 IEEE 20th Conference on Emerging Technologies & Factory Automation (ETFA)*; Luxembourg, Luxembourg, 2015. pp. 1–6.
[26] Lasota P.A., Rossano G.F., Shah J.A. 'Toward safe close-proximity human-robot interaction with standard industrial robots'. *2014 IEEE International Conference on Automation Science and Engineering (CASE)*; New Taipei, Taiwan, 2014. pp. 339–44.
[27] Heyer C. 'Human-robot interaction and future industrial robotics applications'. *2010 IEEE/RSJ International Conference on Intelligent Robots and Systems*; Taipei, Taiwan, 2010. pp. 4749–54.
[28] Laferrière P., Payeur P., Toledo R. 'Instrumented compliant wrist for dexterous robotic interaction'. *2014 IEEE International Symposium on Robotic and Sensors Environments (ROSE)*; Timisoara, Romania, 2014. pp. 66–71.
[29] Das S.K., Saadatzi M.N., Abubakar S., Popa D.O. 'Joint torque estimation using base force-torque sensor to facilitate physical human-robot interaction (pHRI)'. *2019 IEEE 15th International Conference on Automation Science and Engineering (CASE)*; Vancouver, BC, Canada, 2019. pp. 1367–72.
[30] Li T., Zheng K., Liu J., Jia X., Feng J. 'Research on operation intention based on flexible tactile sensing handle'. *IEEE Access*. 2021, vol. 9, pp. 12362–73.

[31] Wang Y., Wu X., Mei D., Zhu L., Chen J. 'Flexible tactile sensor array for distributed tactile sensing and slip detection in robotic hand grasping'. *Sensors and Actuators A*. 2019, vol. 297, p. 111512. Available from https://www.sciencedirect.com/science/article/pii/S0924424718318582

[32] Maggiali M., Cannata G., Maiolino P., Metta G., Randazzo M., Sandini G. 'Embedded distributed capacitive tactile sensor'. *Mechatronics*. 2008, vol. 07.

[33] Yu P., Liu W., Gu C., Cheng X., Fu X. 'Flexible piezoelectric tactile sensor array for dynamic three-axis force measurement'. *Sensors (Basel, Switzerland)*. 2016, vol. 16(6), 819. Available from https://www.mdpi.com/1424-8220/16/6/819

[34] Krishna G.M., Rajanna K. 'Tactile sensor based on piezoelectric resonance'. *IEEE Sensors Journal*. 2004, vol. 4(5), pp. 691–97.

[35] Yuan W., Dong S., Adelson E.H. 'GelSight: high-resolution robot tactile sensors for estimating geometry and force'. *Sensors*. 2017, vol. 17(12), p. 2762. Available from https://www.mdpi.com/1424-8220/17/12/2762

[36] Li W., Konstantinova J., Noh Y., Ma Z., Alomainy A., Althoefer K. 'An elastomer-based flexible optical force and tactile sensor'. *2019 2nd IEEE International Conference on Soft Robotics (RoboSoft)*; Seoul, Korea (South), 2019. pp. 361–66.

[37] Mittendorfer P., Yoshida E., Cheng G. 'Realizing whole-body tactile interactions with a self-organizing, multi-modal artificial skin on a humanoid robot'. *Advanced Robotics*. 2015, vol. 29(1), pp. 51–67. Available from https://doi.org/10.1080/01691864.2014.952493

[38] Wosch T., Feiten W. 'Reactive motion control for human-robot tactile interaction'. *2002 IEEE International Conference on Robotics and Automation (Cat. No.02CH37292)*; Washington, DC, USA, 2002. pp. 3807–12 .

[39] Schmidt P.A., Maël E., Würtz R.P. 'A sensor for dynamic tactile information with applications in human-robot interaction and object exploration'. *Robotics and Autonomous Systems*. 2006, vol. 54(12), pp. 1005–14.

[40] Frigola M., Casals A., Amat J. 'Human-robot interaction based on a sensitive bumper skin'. *2006 IEEE/RSJ International Conference on Intelligent Robots and Systems*; Beijing, China, 2006. pp. 283–87.

[41] Leboutet Q., Dean-Leon E., Cheng G. 'Tactile-based compliance with hierarchical force propagation for omnidirectional mobile manipulators'. *2016 IEEE-RAS 16th International Conference on Humanoid Robots (Humanoids)*; Cancun, IEEE, 2016. pp. 926–31.

[42] Naya F., Yamato J., Shinozawa K. 'Recognizing human touching behaviors using a haptic interface for a pet-robot'. *IEEE SMC'99 Conference Proceedings. 1999 IEEE International Conference on Systems, Man, and Cybernetics (Cat. No.99CH37028)*; Tokyo, Japan, 1999. pp. 1030–34.

[43] Stiehl W.D., Breazeal C. 'Affective computing and intelligent interaction'. *International Conference on Affective Computing and Intelligent Interaction*; Beijing, China, Springer, 2005. pp. 747–54.

[44] Kaboli M., Long A., Cheng G. 'Humanoids learn touch modalities identification via multi-modal robotic skin and robust tactile descriptors'. *Advanced*

Robotics. 2015, vol. 29(21), pp. 1411–25. Available from https://doi.org/10.1080/01691864.2015.1095652

[45] Albini A., Denei S., Cannata G. 'Human hand recognition from robotic skin measurements in human-robot physical interactions'. *2017 IEEE/RSJ International Conference on Intelligent Robots and Systems (IROS)*; Vancouver, BC, 2017. pp. 4348–53.

[46] Liu H., Song X., Nanayakkara T., Seneviratne L.D., Althoefer K. 'A computationally fast algorithm for local contact shape and pose classification using a tactile array sensor'. *IEEE International Conference on Robotics and Automation (ICRA)*; Saint Paul, MN, USA, IEEE, 2012. pp. 1410–15.

[47] Liu H., Greco J., Song X., Bimbo J., Seneviratne L., Althoefer K. 'Tactile image based contact shape recognition using neural network'. *2012 IEEE International Conference on Multisensor Fusion and Integration for Intelligent Systems (MFI 2012)*; Hamburg, Germany, IEEE, 2012. pp. 138–43.

[48] Cao L., Kotagiri R., Sun F., Li H., Huang W., Aye Z.M.M. 'Efficient spatio-temporal tactile object recognition with randomized tiling convolutional networks in a hierarchical fusion strategy'. *Proceedings of the AAAI Conference on Artificial Intelligence*. 2016, vol. 30(1).

[49] Gandarias J.M., Gómez-de-Gabriel J.M., García-Cerezo A.J. 'Enhancing perception with tactile object recognition in adaptive grippers for human-robot interaction'. *Sensors (Basel, Switzerland)*. 2018, vol. 18(3), 692.

[50] Kong Y.-K., Lowe B.D. 'Optimal cylindrical handle diameter for grip force tasks'. *International Journal of Industrial Ergonomics*. 2005, vol. 35(6), pp. 495–507.

[51] Cyskin, Intelligent Hub Board. Available from https://www.cyskin.com/ihb-the-intelligent-hub-board/

[52] Albini A., Denei S., Cannata G. 'Human hand recognition from robotic skin measurements in human-robot physical interactions'. *2017 IEEE/RSJ International Conference on Intelligent Robots and Systems (IROS)*; Vancouver, BC, Canada, 2017. pp. 4348–53.

[53] Ioffe S., Szegedy C. 'Batch normalization: accelerating deep network training by reducing internal covariate shift'. *CoRR*. 2015. Available from http://arxiv.org/abs/1502.03167

[54] Agarap A.F. 'Deep learning using rectified linear units (relu)'. *ArXiv Preprint ArXiv:180308375*. 2018.

[55] Yang Y., Peng Y., Pu H., *et al*. 'Deployable parallel lower-mobility manipulators with scissor-like elements'. *Mechanism and Machine Theory*. 2019, vol. 135, pp. 226–50.

[56] Lu S., Zlatanov D., Ding X., Molfino R. 'A new family of deployable mechanisms based on the hoekens linkage'. *Mechanism and Machine Theory*. 2014, vol. 73, pp. 130–53.

[57] Ramadoss V., Zlatanov D., Ding X., Zoppi M., Lyu S. 'Design, construction, and control of curves and surfaces via deployable mechanisms'. *Journal of Mechanisms and Robotics*. 2019, vol. 11(6).

[58] Dai J.S., Jones J.R. 'Mobility in metamorphic mechanisms of foldable/erectable kinds'. *ASME 1998 Design Engineering Technical Conferences*; Atlanta, GA, USA, 1999.

[59] Gibson C.G., Marsh D. 'On the linkage varieties of watt 6-bar mechanisms—I. basic geometry'. *Mechanism and Machine Theory*. 1989, vol. 24(2), pp. 105–13.

Chapter 8

Online cooperative task execution through human motion capture databases

Miha Deniša[1] and Aleš Ude[1]

This chapter addresses cooperative task executions using a hierarchical database of example human motions. The approach is based on a dual database of example movements, which are gained through demonstrations of human-robot collaborative tasks. Example human movements are encoded in the primary database, which is enhanced with weighted directed graphs. The primary database is used for online human movement recognition through a hierarchical search. The secondary database encodes corresponding robot movements and is used for cooperative robot trajectory synthesis. Smooth and continuous trajectories are gained by encoding them with dynamic movement primitives. The proposed approach was used to execute dynamic cooperative tasks.

8.1 Introduction

The desire to utilize robots in less structured environments, e.g., small-to-medium enterprises or home environments, is on the rise. As the required robot tasks and conditions are prone to change, programming robot movements by hand is unfeasible. Alternatively, new sensorimotor knowledge can be gained through human demonstration using programming by demonstration [1–3]. Multiple approaches can be used to observe human movement: vision systems based on RGB-D sensors [4] or stereo cameras [5]; optical or magnetic marker-based system [6, 7]; or kinesthetic guidance, where a human physically guides the robot [8, 9].

As is the case with dynamic movement primitives (DMPs) [10, 11], robot trajectories can be learned using a single demonstration. On the other hand, multiple demonstrations can be used to learn more flexible robot control policies. DMPs that can adapt to different task conditions can be learned from a set of recorded movements [12, 13]. Other representations of movement primitives that can be used with

[1]Humanoid and Cognitive Robotics Lab, Department of Automatics, Biocybernetics and Robotics, Jožef Stefan Institute, Ljubljana, Slovenia

multiple demonstrations include Gaussian mixture models [14, 15], probabilistic movement primitives [16], hidden Markov models (HMMs) [17], and so on.

One of the major features needed to enable the introduction of robots into home environments and industry settings of small-to-medium enterprises is human-robot collaboration. Human-robot collaboration often requires a two-step process: recognition of human movements and the generation of appropriate collaborative robot movements. Human-robot collaboration based on HMMs was implemented by multiple authors. A path-map HMM was used by Amor *et al.* [18] to model the interaction between two demonstrators while executing a collaborative task. Responsive robot behaviors were learned using hierarchical HMMs by Lee *et al.* [19]. Ben Amor *et al.* [20, 21] proposed an extension of DMPs, called Interaction Primitives (IPs). Probabilistic movement primitives [22] were also used for collaborative task execution [23, 24]. Mavsar *et al.* proposed to compute collaborative robot movements represented by DMPs with deep neural networks [25]. The work most related to our proposed approach was done by Yamane *et al.* [26, 27]. Similarly to our work, a binary tree database was used to recognize human movement and then synthesize an appropriate robot collaborative movement.

In this chapter, an approach based on an extended hierarchical database (HDb) of demonstrated cooperative movements is proposed. A single HDb was already used to synthesize new movements from a set of example trajectories [28]. New compliant movements were also generated through an extended HDb [29]. While kinematic trajectories were encoded in the primary database, the corresponding torques were stored in the secondary part of the HDb. An evaluation of human movement recognition based on HDb is provided in Reference 30.

The rest of the chapter is structured as follows. Section 8.2 presents human movement recognition and cooperative robot trajectory synthesis in three subsections. Section 8.3 focuses on the practical implementation of the proposed approach. The concluding remarks are given at the end of the chapter.

8.2 Cooperative task execution

The approach for the execution of cooperative tasks can be roughly divided into three parts. While the first subsection describes HDb construction, the second presents human movement recognition through hierarchical search. The last subsection explains cooperative robot movement prediction and trajectory synthesis. A block diagram presenting an overview of the process can be seen in Figure 8.1. The diagram is explained in detail in the implementation section.

8.2.1 Hierarchical database

The process of building the HDb starts by performing a set of cooperative movements by two demonstrators, where one demonstrator takes the role of a human and the second the role of the robot in the cooperative task. The set of movements performed by the first demonstrator is denoted as a set of human movements, while the second demonstrator performs the corresponding collaborative robot movements (see Figure 8.2). Although

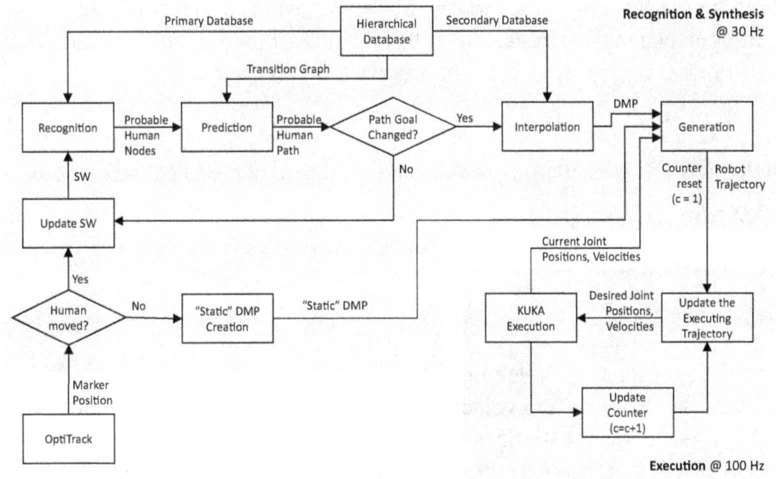

Figure 8.1 *Flowchart depicting the implementation of movement recognition and cooperative task execution*

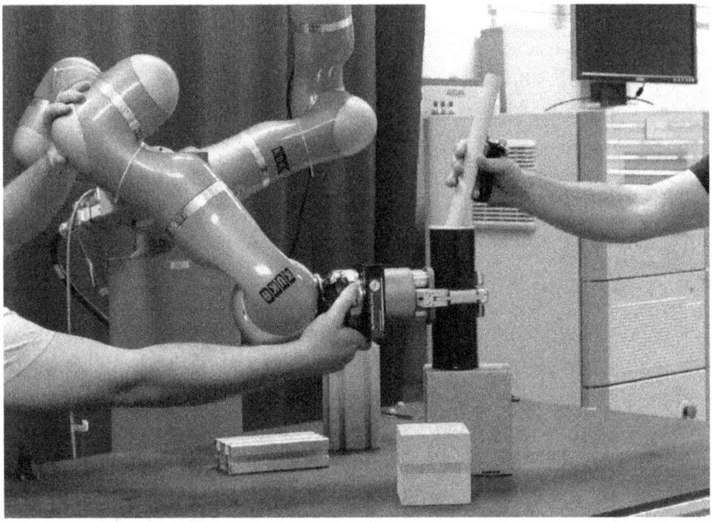

Figure 8.2 *Demonstration of a cooperative task. While one demonstrator executes the human part of the task, the second demonstrator kinesthetically moves the robot in an appropriate manner. The human part of the task is captured using a marker-based vision system.*

multiple approaches can be used to capture these movements, we used a marker-based system for human movements and kinesthetic guidance for robot movements. Through demonstrations, a set of n_S pairs of cooperative movements is captured:

$$\mathcal{S}_D = \{\boldsymbol{D}_1, \boldsymbol{D}_2, \ldots, \boldsymbol{D}_{n_S}\} \tag{8.1}$$

Each demonstration contains n_i state vectors sampled at the fixed time step Δt:

$$\boldsymbol{D}_i = [\boldsymbol{y}_{i,1}, \boldsymbol{y}_{i,2}, \ldots, \boldsymbol{y}_{i,n_i}] \tag{8.2}$$

where the state vectors

$$\boldsymbol{y}_{i,j} = [\boldsymbol{y}_{i,j}^{hT}, \boldsymbol{y}_{i,j}^{rT}]^T, \quad j = 1, \ldots n_i \tag{8.3}$$

include the positions and velocities $\boldsymbol{y}_{i,j}^h$ belonging to the human trajectories, while $\boldsymbol{y}_{i,j}^r$ include the positions and velocities from the corresponding robot movement. If the set consists of demonstrations with different durations, then the number of state vectors $n_{i,D}$ will vary between them. Both the human and robot state vectors can be defined differently. State vectors for end-effector trajectories can, for example, be specified in the Cartesian space and defined as

$$\boldsymbol{y}_{i,j}^h = \left[p_{i,j,x}, \dot{p}_{i,j,x}, p_{i,j,y}, \dot{p}_{i,j,y}, p_{i,j,z}, \dot{p}_{i,j,z}\right]^T \tag{8.4}$$

and denotes the position and velocity at time $t_{i,j}$. If the robot trajectories are specified in joint space, then state vectors are defined as follows:

$$\boldsymbol{y}_{i,j}^r = \left[q_{i,j,1}, \dot{q}_{i,j,1}, q_{i,j,2}, \dot{q}_{i,j,2}, \ldots, q_{i,j,\mathrm{DOF}}, \dot{q}_{i,j,\mathrm{DOF}}\right]^T \tag{8.5}$$

where the kth joint angle and its velocity at time $t_{i,j}$ are denoted by $q_{i,j,k}$ and $\dot{q}_{i,j,k}$, respectively, and DOF stands for the number of degrees of freedom.

The next step in our system is to encode the set $\mathcal{S}_D$ in an extended HDb, where the human set is used to build the primary database and the robot set is encoded in the secondary database. A simple representation of an example HDb can be seen in Figure 8.3. In the first step, all human state vectors $\boldsymbol{y}^h$ belonging to all of the demonstrated movements $\boldsymbol{D}$ are concatenated in a human sample motion matrix:

$$\boldsymbol{Y}^h = [\boldsymbol{y}_1^h, \boldsymbol{y}_2^h, \ldots, \boldsymbol{y}_{n_Y}^h] \tag{8.6}$$

where n_Y denotes the number of state vectors in all demonstrations. This human sample motion matrix represents the root node. The next levels are gained by the division of nodes at the previous layer. Division is done by clustering each node's state vectors into two clusters. While each node could be divided into more than two clusters and could vary from level to level, by selecting two, the level of resolution of the represented data changes gradually throughout the levels of the database. The k-means clustering algorithm is used, as it proved to be the best suited for building an HDb [28]. The division of nodes continues at each level until all nodes meet a *stopping criterion*, which is based on the variability of state vectors associated with the associated node. By copying leaf nodes, each branch is extended to the last level, and thus all the data are represented at each level. For each node $\boldsymbol{v}$, the mean of the associated state vectors $\bar{\boldsymbol{y}}_v^h$ is calculated and stored.

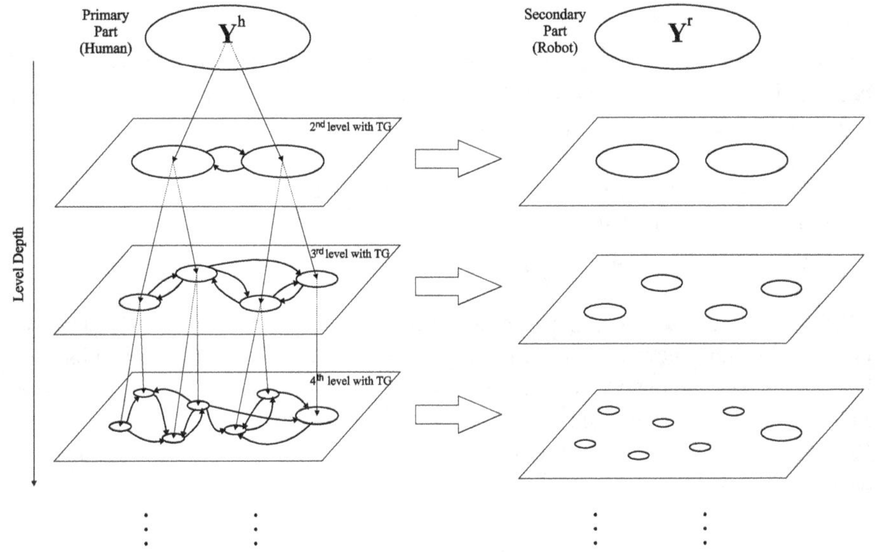

Figure 8.3 *A simple representation of an example HDb. The primary part of the database can be seen on the left. It is used for construction and motion recognition and stores human movements. The secondary part, which encodes the corresponding robot movements, can be seen on the right. Recognition results are used to determine the most probable robot movement, as denoted by hollow arrows.*

All the demonstrated human data are clustered in the process. This means that the time component and the demonstrated sequence of state vectors are lost. The original state vector sequence is instead represented through directed weighted graphs. The transition graph (TG), which represents all possible transitions between the nodes at a certain level, is therefore constructed at each level of the HDb (see Figure 8.3). The weights of the TG's edges represent the probability of transition from one node to another. The edge weights are based on the number of transitions between the nodes observed in the originally demonstrated data. Further details on HDb construction are omitted, and the reader is referred to References 28 and 29.

The primary part of the database only encodes the human state vectors and is used for human movement recognition. While possible to cluster both human and robot data in a single tree, robot data would diminish the performance of human movement recognition. To include the robot state vectors of the associated collaborative movement, the secondary database is constructed by *copying* the nodes from the primary database. This means that for each node in the primary database, a node is created in the secondary one. The nodes encode robot state vectors captured at the same time as the human state vectors in the corresponding node. The mean of the robot state vectors $\bar{y}^r_v$ is calculated and stored for each node $\boldsymbol{v}$. A hierarchical representation example of the database is shown in Figure 8.3.

8.2.2 Human movement recognition

Before an appropriate cooperative robot motion can be generated, the human movement needs to be recognized. To do that, the most probable sequence of nodes in the HDb needs to be determined. A sliding window (SW) of current human motions and hierarchical search through the primary database is used for this purpose.

In the first step, the SW is updated, i.e., the sequence of last n_W human state vectors observed in the current captured human movement:

$$\boldsymbol{Y}^o = \{\boldsymbol{y}^o_{no-n_W}, \ldots, \boldsymbol{y}^o_{no-1}, \boldsymbol{y}^o_{no}\} \tag{8.7}$$

where $\boldsymbol{y}^o$ denotes the *observed* human state vectors, while n_o is the complete number of observed human state vectors from the start of the movement. Naturally, the observed human movements contain only human state vectors $\boldsymbol{y}^h$ [see (8.3)]. The size of the SW n_W is a compromise between the confidence of the result and the speed of recognition. Recognition is done by traversing through the levels of the basic HDb. Multiple steps are carried out at each level l:

1. Establish the considered nodes $\boldsymbol{v}^c$ at current level l. At the first level of the HDb used for recognition (usually level 3), all nodes are marked as considered nodes.
2. A matrix of the considered nodes at current level $\boldsymbol{G}^l$ is built, where each row contains a permutation of considered nodes $\boldsymbol{v}^c$ and has the length of the SW n_W. Matrix $\boldsymbol{G}^l$ includes all possible permutations.
3. The recognition score for each permutation set is calculated:

$$R = K_d \sum_{i=1}^{n_W} d(\boldsymbol{y}^o_i, \boldsymbol{v}^c_i) + \sum_{i=1}^{n_W-1} \tau(\boldsymbol{v}^c_i, \boldsymbol{v}^c_{i+1}) \tag{8.8}$$

   where $d(\boldsymbol{y}^o_i, \boldsymbol{v}^c_i)$ denotes the Euclidean distance between the observed human state vector $\boldsymbol{y}^o_i$ and the mean state vector corresponding to the considered node $\boldsymbol{v}^c_i$. The distance weighting factor K_d is used to weight the distance term in the overall score. It must be established empirically. The second term, $\tau(\boldsymbol{v}^c_i, \boldsymbol{v}^c_{i+1})$, denotes the transition probability between the considered nodes, which is derived from the TG at the current level.
4. Determine the nodes belonging to permutations with the recognition score R above the cut-off range. The cut-off range governs the trade-off between the speed of recognition and the confidence of the result. Only the children of these nodes are considered as we move to the next level of the HDb.

With the last level reached, the permutation of nodes with the highest recognition score is the recognized *human sequence*, i.e., the most probable sequence of nodes with respect to the observed human movement in the current SW.

8.2.3 Cooperative movement prediction and synthesis

Once the most probable human sequence has been determined, the prediction of further human movement can be estimated by extrapolating from the sequence.

By continuing from the found sequence through the corresponding TG, following the edges with the highest transition probability, the most probable subsequent sequence of human nodes can be inferred. The highest weighted edges are followed until one of the end nodes. We denote this sequence as a *human path.*

The most probable human path is found by traversing the primary part of HDb. It is then copied onto the secondary part of the HDb which encodes the cooperative robot movements. This corresponding sequence of nodes is denoted as a *robot path.* In the process of building the HDb, the time component of demonstrations was lost. In order to enhance the found robot path with time stamps, a duration t_v of a single node v is estimated as

$$t_v = \frac{n_v}{m_v}\Delta t \tag{8.9}$$

where m_v denotes the number of demonstrated trajectories passing through the node v and n_v denotes the number of state vectors clustered in it. By combining these estimated node durations with state vectors means $\bar{y}^r_v$, the most probable robot movement can be written as a sequence:

$$\{(\bar{y}^r_1, T_1), (\bar{y}^r_2, T_2), \ldots, (\bar{y}^r_{nR}, T_{nR})\}, \tag{8.10}$$

where n_R denotes the number of nodes in the *robot path.* Time stamps for each node's mean state vector are defined as

$$T_i = \begin{cases} 0 & i = 1 \\ T_{i-1} + \frac{t_{v_{i-1}} + t_{v_i}}{2} & i > 1 \end{cases} \tag{8.11}$$

In the last step, DMPs are used to interpolate the most probable robot path and encode each dimension of the movement. The initial DMP position and velocity values are set to the current robot values. As the human movement recognition and robot synthesis is done online, a change in the most probable *human path* can cause abrupt variation in the desired robot positions during the execution of the robot movement. By setting the DMP's start values to the current robot values, a continuous and smooth transition can be achieved. The details on DMPs are omitted and the readers are referred to References 10 and 11.

8.3 Experimental evaluation

We have implemented a cooperative task to evaluate the proposed approach. While the human is picking up one of the two objects and moving it to various positions, the robot needs to appropriately turn the object held in its hand and move it to the correct end position. The setup consisted of two Kuka LWR-4 robot arms, a three-finger Barrett gripper holding a hollow cylinder-like object, and a marker-based OptiTrack vision system. One of the robot arms was used for executing the cooperative robot movements. The other arm was used as a mounting point for the vision system, which tracked three markers attached to the human hand. The described setup is shown in Figure 8.2.

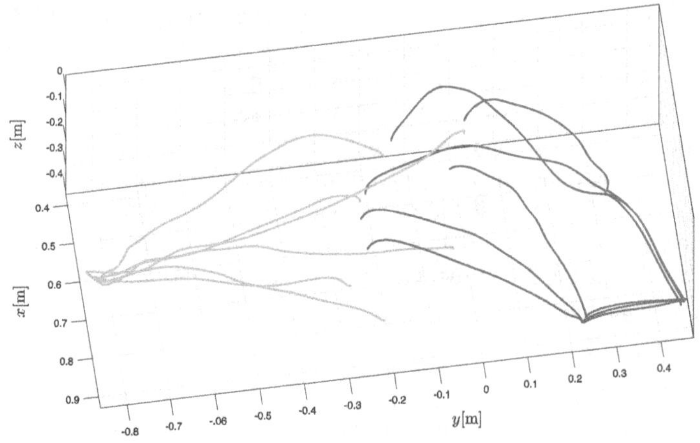

Figure 8.4 *Set of demonstrated cooperative trajectories. Robot trajectories, seen in yellow, were converted to task space for presentation purposes. Human trajectories are denoted in blue.*

The setup was used to capture a set of six cooperative movements:

$$\mathcal{S}_Y = \{\boldsymbol{Y}_1, \boldsymbol{Y}_2, \ldots, \boldsymbol{Y}_6\} \tag{8.12}$$

consisting of human and robot trajectories. While one demonstrator took the role of the robot and guided it kinesthetically, the other performed human movements while having markers attached to the hand. The demonstration process is shown in Figure 8.2. While the combination of marker-based and kinesthetic demonstration was used in our experiment, other techniques could be used. Kinesthetic guidance of the robot's movements has the advantage of gaining movements directly applicable to the robot.

The human trajectories consisted of positional state vectors:

$$y^h = \left[p_x^h, \dot{p}_x^h, p_y^h, \dot{p}_y^h, p_z^h, \dot{p}_z^h\right]^T \tag{8.13}$$

while cooperative robot movements were captured in joint space, with state vectors defined as

$$y^r = \left[q_1^r, \dot{q}_1^r, q_2^r, \dot{q}_2^r, \ldots, q_7^r, \dot{q}_7^r\right]^T \tag{8.14}$$

The set of demonstrated cooperative trajectories can be seen in Figure 8.4, where the robot trajectories were converted to task space for presentational purposes. As can be seen in Figure 8.4, the set of six cooperative movements can be roughly divided into two subgroups. Each subgroup, consisting of three cooperative movements, executes a variation of a pick-and-place task. The human had to pick up one of the two objects, move it to one of three positions (varying in all three dimensions), and insert/put the object in/on the cylinder, which the robot is holding. If the human picks up the wooden peg, seen in Figure 8.2, he moves it to one of the leftmost positions and then inserts it into the cylinder. If he picks up the plastic cover, he moves it to one of the right-most positions and put it on the side of the cylinder with no opening. The demonstrator who guides the cooperative robot moves it to the proper final position and rotates the cylinder to the correct rotation.

The set of demonstrated movements was used to build the extended HDb, as explained in section 8.2.1. The HDb was then used for online movement recognition and cooperative task execution. While the basic principle explained in sections 8.2.2 and 8.2.3 was implemented, some further explanation regarding online cooperative task execution is required. The flowchart presented in Figure 8.1 depicts the implementation in some more detail. It can be roughly divided into two main parts: (a) human movement recognition and cooperative trajectory synthesis, which runs at 30 Hz, and (b) execution of the trajectory on the Kuka robot arm, running at 100 Hz.

The OptiTrack camera provides marker positions as input for the recognition and synthesis part. As we do not want the recognition to start before the human starts executing his motion, the first block compares the current human position to the average of all starting positions of the demonstrated human trajectories. We assume that all demonstrated movements start in approximately the same position. As seen in Figure 8.4, this is true for our set of demonstrated movements. If the human hand is in the starting position, a *static* DMP is generated and sent to the Generation block. With static DMP, we denote a DMP, which encodes a trajectory of constant values of mean human starting positions. If the human moves from the start position, marker positions are used to update the SW. The SW consists of the last n_W human position, sampled at 30 Hz. Throughout our experiments, $n_W = 4$ was used. The SW is then sent to the Recognition block, which applies the hierarchical search to the primary database. Recognition, as well as other major components, are described in section 8.2. The identified sequence of the most probable human nodes is then used to infer the most probable human path, which is done by using the TG from the HDb. In the next step, we check if the path goal has changed with respect to the previous recognition cycle. If the final node on the human path remains the same, the human is still following the same demonstrated movement and we do not need to update the robot trajectory currently being executed. If the final node has changed, the new human path is passed to the Interpolation block. The role of the Interpolation block is time estimation and interpolation using DMP encoding. The required data are provided by the secondary database. The DMP parameters are sent to the Generation block that synthesizes the robot position and velocity trajectories. In addition, it resets the executing trajectory counter to $c = 1$. The trajectories, sampled at 100 Hz, start at the current robot position and velocity, which is captured from the robot. The DMP ensures the smooth and continuous transition to the newly synthesized robot trajectory, which is passed to the second major block of the implementation.

The Execution block updates the current robot trajectory whenever it gets a new one from the Generation block. As the counter updates each iteration, the trajectory is being executed on the robot until the counter is set back to one and the trajectory is updated. If the current trajectory comes to an end before the new one is received, the execution is finished, as the robot has reached the final destination.

Figure 8.5 presents the computation times needed for the major blocks of the recognition and task synthesis. The times are means over ten different example executions. These executions included the human moving the object to all six final positions,

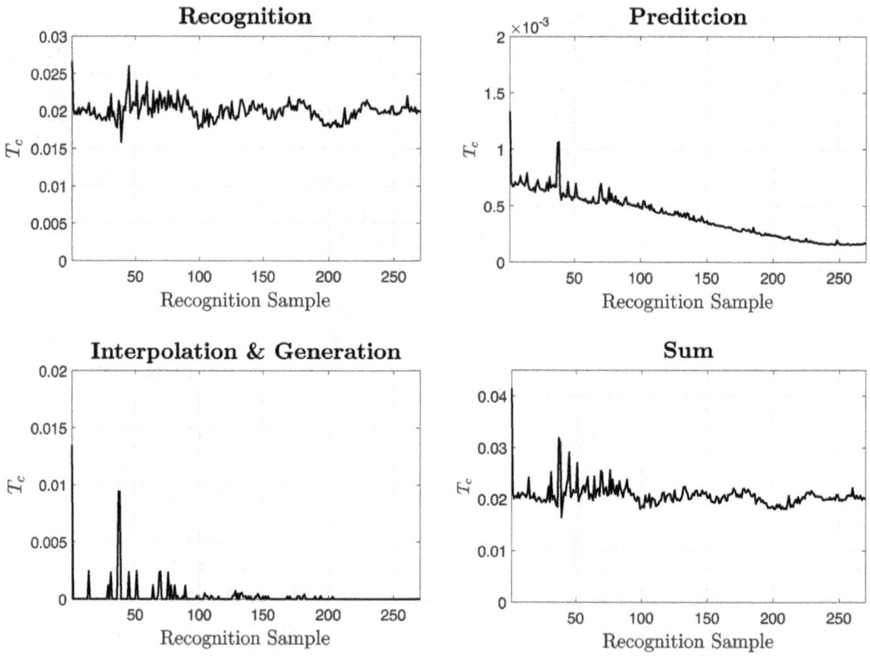

Figure 8.5 *Calculation time needed throughout the experiment. Mean values of calculation times over all 10 executions in the experiment are presented in the four graphs. While the sum is presented in the bottom-right graph, the other three present calculation times needed for separate components of the process, as presented in Figure 8.1.*

in-between changes regarding the picked-up object, and in-between changes regarding the final position. Besides the computation times of the major blocks, the overall computation is also presented in the lower-right graph. We can observe the computation times around 0.02 seconds, with an occasional rise above the desired 0.033 seconds, i.e., 30 Hz.

Video stills from two example executions, seen in Figure 8.6, show the robot executing the appropriate cooperative movements with respect to the human motion. In the first example, the human picks up the cover, while in the second one, he picks up the peg. We can observe that the robot rotates the object accordingly and then moves it to the appropriate final position. Video stills in Figure 8.7 show examples, where the human changes his mind about which object to pick up during the execution. In both examples, the robot acts accordingly with smooth and continuous movements.

8.4 Conclusion

This chapter presents a new approach to the generation of cooperative robot movements based on an HDb of example human motions. Demonstrations are encoded

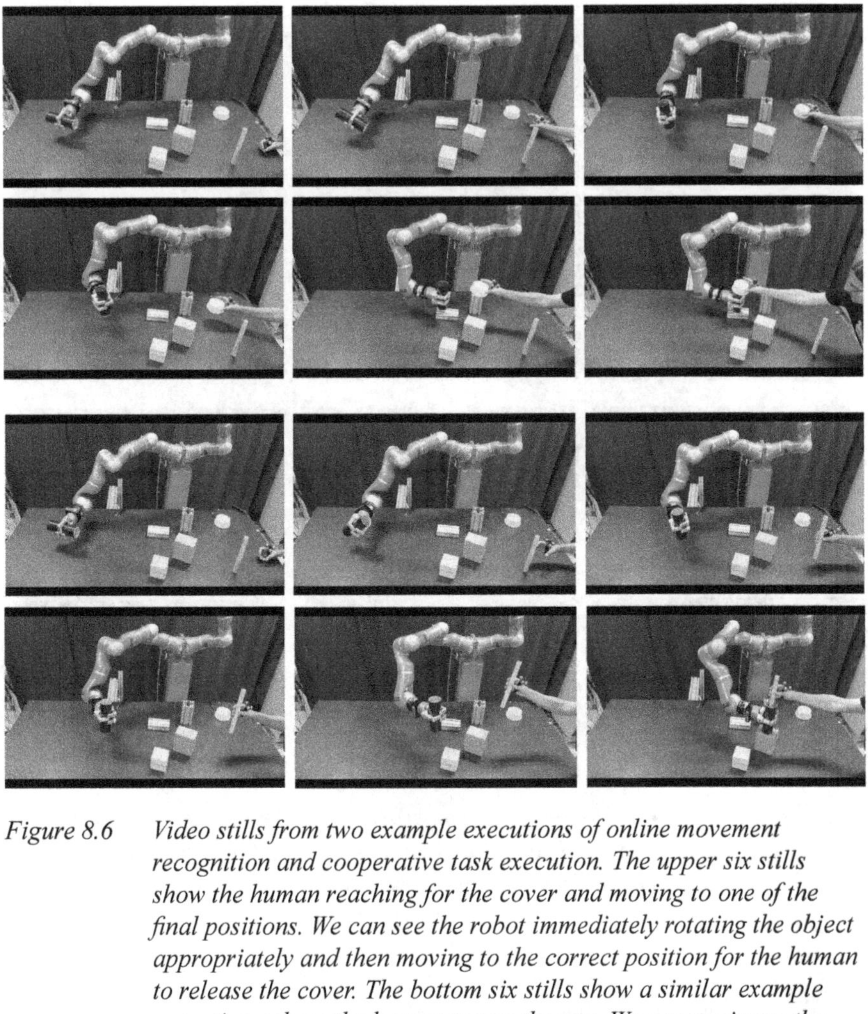

Figure 8.6 *Video stills from two example executions of online movement recognition and cooperative task execution. The upper six stills show the human reaching for the cover and moving to one of the final positions. We can see the robot immediately rotating the object appropriately and then moving to the correct position for the human to release the cover. The bottom six stills show a similar example execution, where the human moves the peg. We can again see the robot rotating and moving the object accordingly.*

in a dual binary tree-like HDb. The primary database, encoding human movements, is used for human movement recognition. The secondary database stores the corresponding robot example movements. Its data are used with DMPs to synthesize smooth and continuous trajectories that successfully accomplish the desired cooperative tasks. The viability of the approach was shown through the implementation of a cooperative task on a real robotic system.

One of the main differences of our approach compared with the work of Yamane *et al.* [26, 27] concerns the stop clustering criterion. While Yamane *et al.* propose a criterion based on the number of nodes in clusters, we also consider the variability of data in clusters [28]. When it comes to recognition, in Yamane *et al.* [27], the most

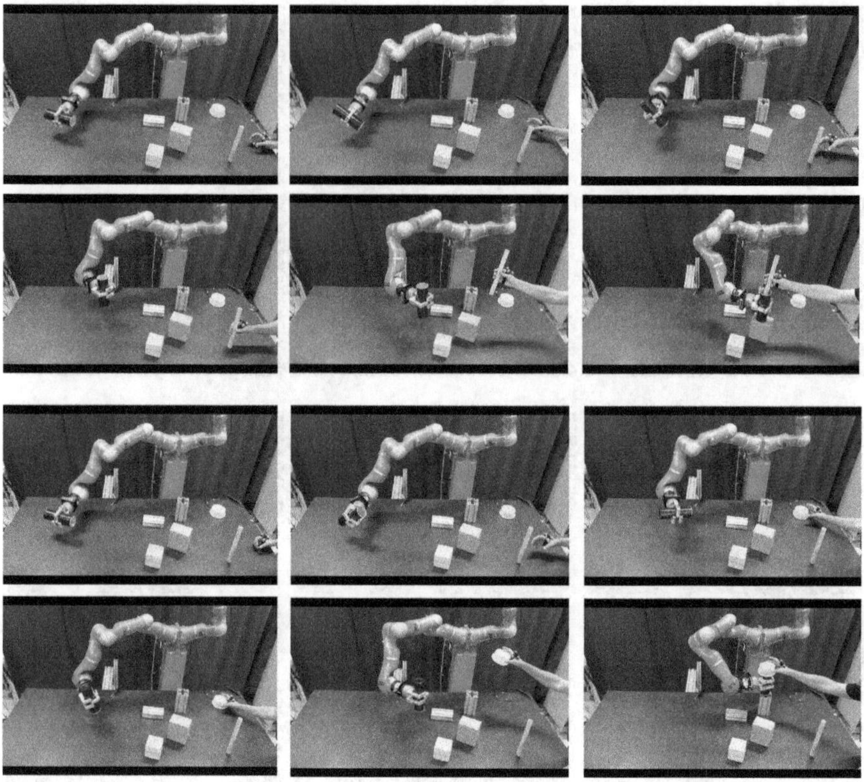

Figure 8.7 *Video stills from two example executions of online movement recognition and cooperative task execution. The upper six stills show the human reaching for the cover and changing the movement mid-execution to pick up the peg. We can see the robot starting the appropriate rotation for the cover, but then transitioning to a newly synthesized cooperative movement when a new human motion is recognized. The bottom six stills show a similar example execution, where the human moves toward the peg and then changes the movement to pick up the cover. We can again see the robot adapting the rotation and movement accordingly.*

probable sequence of nodes is compared with the ones found with respect to the SW in the previous step. This way any changes in human movements with respect to example movements are penalized and sudden changes in the synthesized robot movement are avoided. This means the approach will be slow to change the cooperative movement if the human suddenly changes his motion during recognition. Instead of penalizing changes, we guarantee a smooth and continuous robot trajectory through DMP encoding and its appropriate start values. By traversing the TG, we infer the future human movement and with it the corresponding robot movement.

This enables us to reduce the number of required robot trajectory generations. An extensive comparison to IPs, proposed by Ben Amor *et al.* [20], was performed and presented in Reference 30. Our future work will focus on evaluation with a larger number of versatile example movements.

References

[1] Hussein A., Gaber M.M., Elyan E., Jayne C. 'Imitation learning: a survey of learning methods'. *ACM Computing Surveys*. 2017, vol. 50(2), pp. 1–35.
[2] Ravichandar H., Polydoros A.S., Chernova S., Billard A. 'Recent advances in robot learning from demonstration'. *Annual Review of Control, Robotics, and Autonomous Systems*. 2020, vol. 3(1), pp. 297–330. Available from https://www.annualreviews.org/toc/control/3/1
[3] Dillmann R. 'Teaching and learning of robot tasks via observation of human performance'. *Robotics and Autonomous Systems*. 2004, vol. 47(2–3), pp. 109–16.
[4] Shotton J., Sharp T., Kipman A., *et al.* 'Real-Time human pose recognition in parts from single depth images'. *Communications of the ACM*. 2013, vol. 56(1), pp. 116–24.
[5] Moeslund T.B., Hilton A., Krüger V. 'A survey of advances in vision-based human motion capture and analysis'. *Computer Vision and Image Understanding*. 2006, vol. 104(2–3), pp. 90–126.
[6] Pollard N.S., Hodgins J.K., Riley M.J., Atkeson C.G. 'Adapting human motion for the control of a humanoid robot'. *2002 IEEE International Conference on Robotics and Automation*; Washington, DC, USA, 2002. pp. 1390–97.
[7] Ude A., Atkeson C.G., Riley M. 'Programming full-body movements for humanoid robots by observation'. *Robotics and Autonomous Systems*. 2004, vol. 47(2–3), pp. 93–108.
[8] Hersch M., Guenter F., Calinon S., Billard A. 'Dynamical system modulation for robot learning via kinesthetic demonstrations'. *IEEE Transactions on Robotics*. 2008, vol. 24(6), pp. 1463–67.
[9] Rozo L., Calinon S., Caldwell D., Jimenez P., Torras C. 'Learning collaborative impedance-based robot behaviors'. *Proceedings of the AAAI Conference on Artificial Intelligence*. 2013, vol. 27(1), pp. 1422–28.
[10] Schaal S., Mohajerian P., Ijspeert A. 'Dynamics systems vs. optimal control – a unifying view'. *Progress in Brain Research*. 2007, vol. 165, pp. 425–45.
[11] Ijspeert A.J., Nakanishi J., Hoffmann H., Pastor P., Schaal S. 'Dynamical movement primitives: learning attractor models for motor behaviors'. *Neural Computation*. 2013, vol. 25(2), pp. 328–73.
[12] Ude A., Gams A., Asfour T., Morimoto J. 'Task-Specific generalization of discrete and periodic dynamic movement primitives'. *IEEE Transactions on Robotics*. 2010, vol. 26(5), pp. 800–15.

[13] Pahič R., Ridge B., Gams A., Morimoto J., Ude A. 'Training of deep neural networks for the generation of dynamic movement primitives'. *Neural Networks*. 2020, vol. 127, pp. 121–31.

[14] Calinon S., D'halluin F., Sauser E., Caldwell D., Billard A. 'Learning and reproduction of gestures by imitation'. *IEEE Robotics & Automation Magazine*. 2010, vol. 17(2), pp. 44–54.

[15] Pignat E., Calinon S. 'Bayesian Gaussian mixture model for robotic policy imitation'. *IEEE Robotics and Automation Letters*. 2019, vol. 4(4), pp. 4452–58.

[16] Paraschos A., Daniel C., Peters J., Neumann G. 'Using probabilistic movement primitives in robotics'. *Autonomous Robots*. 2018, vol. 42(3), pp. 529–51.

[17] Inamura T., Toshima I., Tanie H., Nakamura Y. 'Embodied symbol emergence based on mimesis theory'. *The International Journal of Robotics Research*. 2004, vol. 23(4–5), pp. 363–77.

[18] Amor H.B., Vogt D., Ewerton M., *et al*. 'Learning responsive robot behavior by imitation'. *Intelligent Robots and Systems (IROS), 2013 IEEE/RSJ International Conference on*; IEEE, 2013. pp. 3257–64.

[19] Lee D., Ott C., Nakamura Y. 'Mimetic communication model with compliant physical contact in human—humanoid interaction'. *The International Journal of Robotics Research*. 2010, vol. 29(13), pp. 1684–704.

[20] Amor H.B., Neumann G., Kamthe S. 'Interaction primitives for human-robot cooperation tasks'. *Robotics and Automation (ICRA), 2014 IEEE International Conference on*; IEEE, 2014. pp. 2831–37.

[21] Ewerton M., Neumann G., Lioutikov R., Ben Amor H., Peters J., Maeda G. 'Learning multiple collaborative tasks with a mixture of interaction primitives'. *2015 IEEE International Conference on Robotics and Automation (ICRA)*; Seattle, WA, USA, IEEE, 2015. pp. 1535–42.

[22] Paraschos A., Daniel C., Peters J.R., *et al*. 'Probabilistic movement primitives' in Burges C.J.C., Bottou L., Welling M. (eds.). *Advances in neural information processing systems 26*. Curran Associates, Inc; 2013. pp. 2616–24.

[23] Maeda G., Ewerton M., Lioutikov R., Ben Amor H., Peters J., Neumann G. 'Learning interaction for collaborative tasks with probabilistic movement primitives'. *2014 IEEE-RAS 14th International Conference on Humanoid Robots (Humanoids 2014)*; Madrid, Spain, IEEE, 2014. pp. 527–34.

[24] Ewerton M., Neumann G., Lioutikov R., Ben Amor H., Peters J., Maeda G. 'Learning multiple collaborative tasks with a mixture of interaction primitives'. *2015 IEEE International Conference on Robotics and Automation (ICRA)*; Seattle, WA, USA, IEEE, 2015. pp. 1535–42.

[25] Mavsar M., Ridge B., Pahič R., Morimoto J., Ude A. 'Simulation-aided handover prediction from video using recurrent image-to-motion networks'. *IEEE Transactions on Neural Networks and Learning Systems*. 2022.

[26] Yamane K., Yamaguchi Y., Nakamura Y. 'Human motion database with a binary tree and node transition graphs'. *Autonomous Robots*. 2011, vol. 30(1), pp. 87–98.

[27] Yamane K., Revfi M., Asfour T. 'Synthesizing object receiving motions of humanoid robots with human motion database'. *Proceedings of IEEE International Conference on Robotics and Automation (ICRA)*; Karlsruhe, Germany, 2013. pp. 1629–36.

[28] Deniša M., Ude A. 'Synthesis of new dynamic movement primitives through search in a hierarchical database of example movements'. *International Journal of Advanced Robotic Systems*. 2015, vol. 12(10), p. 137.

[29] Deniša M., Petrič T., Asfour T., Ude A. 'Synthesizing compliant reaching movements by searching a database of example trajectories'. *Proceedings of IEEE International Conference on Humanoid Robots*; Atlanta, GA, USA, 2013. pp. 540–43.

[30] Deniša M., Nemec B., Ude A. 'Cooperative movements through hierarchical database search'. *Advanced Robotics (ICAR), 2017 18th International Conference on*; IEEE, 2017. pp. 40–46.

Chapter 9

Versatile automation of dense packing with teaching by demonstration and 3D vision

Sotiris Antonakoudis[1], *Nikos Kaparinos*[1], *Antonis Sidiropoulos*[1], *Savvas Sampaziotis*[1], *Marios Kiatos*[1], *Fotios Dimeas*[1], *and Zoe Doulgeri*[1]

Recent advances in robotics have paved the way for the automation of many tasks in the manufacturing industry that were previously performed manually. In particular, through the use of collaborative robots, there are many examples of exciting new applications where humans and robots can work together safely and effectively. However, end-of-line automation, where products are packed into cartons and cartons are placed on pallets, is still done using traditional, time-consuming methods of programming the robot's path and calculating the packing plan. Most importantly, such methods do not incorporate perception and do not take full advantage of a robot's capabilities for a flexible automation cell. For example, in a manufacturing line with frequent batch production changes, packing a different product requires reprogramming of the robot.

In this chapter, we propose a versatile solution for automating dense packing tasks in an industrial environment using teaching by demonstration and 3D vision. Teaching enables the user to create the robot's pick-and-place motion pattern in a very short time and without any programming effort. 3D vision with AI gives the robot the ability to recognize its environment, adapt its movements, and remain safe for humans. The system is able to accurately detect and track objects moving on a conveyor by proposing a novel vision pipeline that has high accuracy and can handle partial or complete occlusion of objects by the robot's body. In addition, the system can detect humans in the robot's workspace and adjust the robot's speed to maintain the minimum distance between the robot arm and the human body. We validate the system in a series of experiments with a dense packing task. Our goal is a portable and flexible robotic cell that can be taught by personnel on the production line how to pack products quickly and easily.

[1]Automation & Robotics Lab, Aristotle University of Thessaloniki, Greece

9.1 Introduction

The most common application for robot arms in the manufacturing industry is object handling.* Large-sized productions usually do not require versatility from robots, as they can dedicate a production line to a single product. However, smaller productions can have multiple product variations on the same production line in a single day. Handling product variations requires enhanced perception capabilities that increase the cost of the entire robotic system due to the additional programming and integration effort. In addition, there is an ongoing trend toward product customization that is forcing even large productions to move to high-mix, low-volume manufacturing. Despite the continuous technological advances aimed at making robots more versatile and user-friendly, they have not yet gained widespread acceptance among small- and medium-sized industries.

Recent research in object handling has focused on how to place objects of different sizes into a container, such as a cardboard box. This task is also known as secondary packaging for end-of-line automation. The authors in Reference 1 proposed a method for a robot to place objects in a confined space, such as a small shipping box, in close vicinity to each other, in an orderly manner and well aligned with respect to the boundaries of the box. The robot can pick up static objects from a pile and, if necessary, topple them with a suction gripper before placing them in the box. While this method provides good robustness to perception errors, it does not handle moving objects or occlusions and relies on automatic motion planners such as MoveIt [2]. In Reference 3, a method for packing objects of different, non-convex shapes is proposed (mixed packing) that ensures the stability of the object stack during packing and closes the loop with vision so that a robot can dynamically react to placement errors and correct them. The same authors have proposed a method for packing objects with known dimensions but in a random arrival order [4]. In Reference 5, packing objects of different sizes are proposed by using Deep Learning to determine the packing strategy. In Reference 6, the authors propose the automatic measurement of object dimensions using an RGB-D camera and optimize their position in a container based on heuristic rules.

The trajectory that the robot executes to perform a handling task is traditionally programmed with online or offline methods, using the teaching pendant of the robot. This is a time-consuming method that also lacks the flexibility to scale the programmed trajectories, for example, to the target placement locations generated by the packaging planners mentioned above. Teaching by demonstration is a promising approach to significantly reduce programming time by transferring skills from a human to a robot [7]. Teaching can be done with kinesthetic guidance, where the human guides the robot through direct manipulation and demonstrates the trajectory. Many of the latest collaborative robots already feature teaching by demonstration for simple via point teaching or for record-and-play of the demonstrated path. While these methods can speed up programming, they are not parametric and offer limited flexibility. On the contrary,

*IFR report: World Robotics 2022 – Industrial Robots

by encoding the demonstrated path with parametric methods, the robot can execute it in a more flexible and adaptable manner. Dynamic movement primitives (DMP) [8] is a parametric method that provides the necessary encoding capabilities of the demonstrated trajectory for flexibility and robustness during execution. It is particularly popular because it provides the ability to scale the demonstrated trajectory temporally and spatially. Recent advances in DMP have enabled new capabilities such as reversible execution of the demonstrated trajectory [9] and online optimization to account for both temporal and spatial constraints [10]. Besides DMP, many methods have been proposed in the literature to achieve these properties, such as dynamical systems [11], Gaussian mixture models [12], and probabilistic movement primitives [13]. The latter methods are more suitable when complex behavioral landscapes from multiple demonstrations need to be encoded.

To achieve maximum versatility of the robotic cell, the system must accurately recognize the pose of target objects without relying on mechanical guides or hard automation, for example, to place each object in a predefined position that is fixed into the robot program. Recent advances in AI have enabled the accurate detection of objects from an RGB image and the estimation of their 6D pose (position and orientation) from the depth image of RGB-D cameras for the robot to pick [14]. Models that are based on Deep Learning form the state of the art in computer vision for object detection. They typically use a backbone network to encode the input image into image features (getting the feature map representation) and a head network to perform the actual detection. The backbone is either a convolution network [15–17] or an attention-based network [18–20], while the head is usually a fully connected one. A common choice for backbone networks is residual convolutional networks (ResNets) [16]. Residual networks use residual connections between layers that allow gradients to propagate through the network without vanishing. Models that are based on Deep Learning can be divided into two categories: single-stage and two-stage. Compared to single-stage, the two-stage models include an additional step to generate a region of interest that may contain the desired object. Two-stage models provide higher performance, especially in detecting small objects. However, the performance gain comes at the cost of increased inference time [21, 22]. A widely used family of two-stage models is the R-CNN. This family includes the R-CNN [23], the Faster R-CNN [24] which is designed to improve computation time, and the Mask R-CNN [25], which, in addition to the other models, also performs instance segmentation (locating the boundaries of each object). While these methods are suitable for detecting the presence of target objects in an image, further steps are required to determine their pose so that they can be picked by a robot.

Methods for 6D pose estimation range from template matching [26] and optimization [27] to Deep Learning [28, 29]. However, they are computationally intensive and take a long time to provide an estimate. This can lead to incorrect estimates if there is relative motion between the camera and the target object. To this end, they can be used in combination with iterative methods to refine the pose and achieve better accuracy. Pose refinement requires an initial rough estimate of the object pose and is usually performed at a higher frequency than pose estimation. RePOSE [30] uses a convolutional neural network along with deep texture rendering to iteratively refine the input pose estimate. RNNPose [31] uses a recurrent neural network to iteratively refine the correspondence field and the pose estimate. Similarly, the coupled iterative refinement [32] uses a novel neural network layer for iterative refinement of the pose estimate. Finally, the iterative corresponding geometry (ICG) method computes the contour points, correspondence lines, and correspondence points between two consecutive images, using optimization to refine the input pose. Pose refinement techniques, such as those mentioned above, provide performance gains at minimal additional computational cost.

Object detection and tracking methods that rely on machine vision are susceptible to occlusions. The robot itself may conceal targets of interest from the camera while moving in the field of view. An error in perception can seriously affect the robot's performance and prevent it from successfully completing its task. There is also a risk that it could be damaged by potential misbehavior. A camera-in-hand configuration is not always practical, for example, when the entire working area must be covered or the measurement distance is less than the minimum allowable distance of the depth sensor. Occlusion detection has been studied in the literature to improve the performance of object-tracking perception systems [33–35] targeting object manipulation. There are also methods specifically for occlusions during human tracking to improve the safety aspect of human-robot collaboration in shared workspaces [36–38].

Information about the structure of the robot, such as the unified robot description format (URDF), has been used in the literature to improve perception. The URDF filtering [39] can be used to subtract the robot model from the depth image. However, this method does not distinguish whether the targets are actually occluded or whether they are between the line of sight from the camera to the robot. To solve this problem, the authors in Reference 40 fused the depth images from two separate sensors. Aside from the cost of the additional sensor, the robot model had to be rendered twice from the perspective of each camera and then fused, which increased the computational cost. In our previous work [41], we proposed a computationally lightweight method to determine the areas that the robot occludes from the camera in real time using a single RGB-D camera. Instead of subtracting the robot model from the depth image, we generate a virtual depth image (VDI) from the URDF model with minimal computational resources. This VDI is fused with the depth image, which is captured by the camera sensor, to determine the area that the robot actually occludes.

A robot that is hand guided and can operate autonomously in an environment without protective fences in close proximity to humans must be safe for human-robot interaction. The first step in avoiding harm to humans is through the use of collaborative manipulators, which have rounded edges and force limitations [42]. However, to ensure that human-robot interaction is performed both safely and effectively, more advanced safety techniques should be employed [43]. These techniques should prevent potential collisions through obstacle avoidance methods [44] and, when collisions are detected, trigger appropriate reaction movements or introduce compliance in the manipulator to avoid high contact forces [45, 46]. The current technical specification ISO/TS 15066 defines different modes of operation to ensure safety in human-robot collaboration. When human-robot contact is required, for example, teaching by demonstration, the hand guidance mode is used. This mode is activated with an enabling device to allow hand guidance of the robot only for intended contacts. For autonomous operation near humans, the speed and separation monitoring mode is used. In this mode, the robot's speed depends on the distance from the nearest human. In Reference 47, the authors used an RGB-D camera to continuously detect human workers in a shared workspace. Skeleton tracking algorithms such as OpenPose or AlphaPose [48] can be used to determine the positions of human skeletal joints and calculate the shortest distance to the robot. Nevertheless, the robustness of such methods has not yet been proven for industrial-grade safety.

In this work, we propose a versatile solution for automating dense packing tasks in an industrial environment based on teaching by demonstration and machine vision. Teaching enables the user to create the robot's pick-and-place motion pattern in a very short time with zero-code programming. Machine vision with AI provides the robot with the ability to recognize its environment, adapt its movements, and remain safe for human interaction. The system is able to accurately detect and track objects moving on a conveyor by proposing a novel machine vision pipeline that has high accuracy and can handle partial or full occlusions of objects by the robot body. In addition, the system can detect humans within the robot's workspace and adjust the robot's speed to maintain the minimum distance between the robot arm and the human body. We validate the system in a series of experiments with a dense packing task. Our goal is a portable and flexible work cell with a robotic arm that can be placed on the production line and be taught very easily and quickly by personnel how to pack products.

The remaining of the chapter is structured as follows. In section 9.2, we present the overall architecture of the system which consists of three subsystems. Then, each subsystem is presented: the packing controller in section 9.3, the object perception in section 9.4, and the human perception in section 9.5. Finally, the experimental validation of the system is presented in section 9.6, followed by the conclusions of the chapter in section 9.7.

9.2 System architecture for a versatile and safe packing system

The proposed system that is shown in Figure 9.1 consists of a collaborative robot with a suction end-effector. A conveyor belt feeds objects to the work area where the packing takes place. An overhead camera above the conveyor identifies the incoming objects. Another camera monitors the entire robot workspace to avoid collisions with humans. The key features we aim for are ease of use, versatility, and safety. As shown in the architecture block diagram in Figure 9.2, there are three main subsystems:

Packing controller. It is the key subsystem that controls the motion of the robot to allow teaching by demonstration, motion encoding, and motion generation for autonomous execution. It provides a compliant behavior of the robot for the demonstration and encodes the path using parametric methods based on DMP. The learned trajectory can be then scaled to new pick locations provided by the object perception

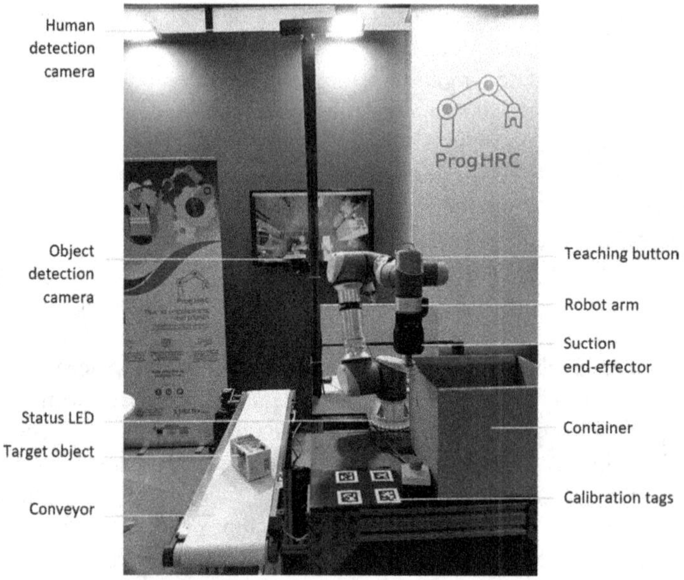

Figure 9.1 System setup for dense packing

system. This subsystem also calculates the packing plan according to the first demonstration to also scale the trajectory with respect to the place position.

Object perception. Detects and tracks the objects to be picked by the robot. This subsystem provides the needed versatility using machine vision to observe the robot's workspace through an RGB-D camera to identify the objects it needs to pick while they are moving. Estimating the position and orientation of objects in space will feed the dynamic motion generation system that will be able to automatically adapt and guide the robot to grasp objects in new positions without the need for human intervention.

Human perception. This subsystem is responsible for the safety of the humans around the robot as it provides their location with respect to the robot. For human recognition, an AI method is used to extract the human skeleton from an RGB-D camera. Finally, the subsystem calculates the minimum distance of the robot from

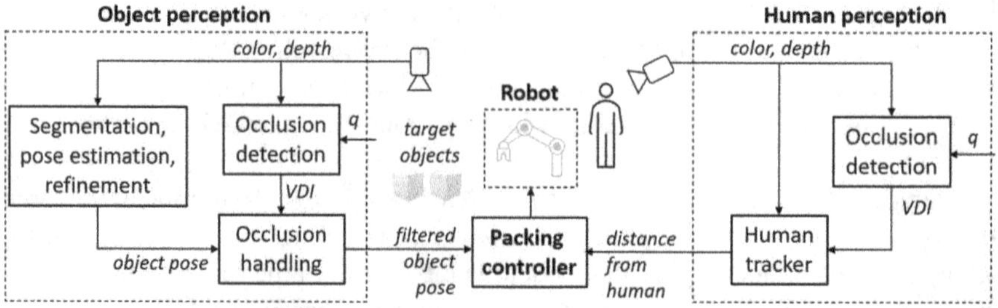

Figure 9.2 Block diagram of the overall architecture

the humans and then adjusts the speed of the end-effector according to the safety standards.

9.3 Packing controller

Let $\mathbf{r}(t) = \{\mathbf{p}, \mathbf{Q}\}$ be the task coordinates of the robot in the task space, where $\mathbf{p} \in \mathbb{R}^3$ is the position and $\mathbf{Q} \in \mathbb{S}^3$ the orientation expressed as a unit quaternion ($\mathbb{S}^3$ is a unit sphere in $\mathbb{R}^4$). During kinesthetic teaching, the robot should be compliant to external forces by the operator and to record the task coordinates $\mathbf{r}(t)$. After teaching, the recorded path is encoded, and the DMP-based motion generation system (presented in the following subsections) generates the desired trajectory $\mathbf{r}_d(t) = \{\mathbf{p}_d, \mathbf{Q}_d\}$ to be followed by the robot during execution.

9.3.1 Admittance controller

Kinesthetic teaching and autonomous execution are achieved using the same admittance controller. This controller renders the following dynamic behavior of a mass-spring-damper system at the end-effector based on the measured interaction forces:

$$\mathbf{M}\dot{\tilde{\mathbf{v}}} + \mathbf{C}\tilde{\mathbf{v}} + \mathbf{K}\tilde{\mathbf{x}} = \mathbf{F}_h \tag{9.1}$$

where $\mathbf{M}$, $\mathbf{C}$, $\mathbf{K} \in \mathbb{R}^{6\times 6}$ are the parameters for desired inertia, damping, and stiffness, respectively. $\mathbf{F}_h \in \mathbb{R}^6$ is the external force measured by a wrist force/torque sensor. $\tilde{\mathbf{x}} \in \mathbb{R}^6$ and $\tilde{\mathbf{v}} \in \mathbb{R}^6$ are the tracking pose, and velocity errors are described as

$$\tilde{\mathbf{x}} = \begin{bmatrix} \mathbf{p} - \mathbf{p}_d \\ 2\log\left(\mathbf{Q} * \mathbf{Q}_d^{-1}\right) \end{bmatrix}, \quad \tilde{\mathbf{v}} = \begin{bmatrix} \dot{\mathbf{p}} - \dot{\mathbf{p}}_d \\ \boldsymbol{\omega} - \boldsymbol{\omega}_d \end{bmatrix}$$

The orientation error is the quaternion log of the quaternion product $*$ between the current orientation $\mathbf{Q}$ of the robot end-effector and the conjugate of the desired one $\mathbf{Q}_d$. The variables $\boldsymbol{\omega}$ and $\boldsymbol{\omega}_d$ are the actual and desired angular velocities of the end-effector.

For the teaching stage, the stiffness term $\mathbf{K}$ is set to zero to allow hand guidance. Since there is no desired trajectory yet, the desired variables for position $\mathbf{p}_d$ and angular velocity $\boldsymbol{\omega}_d$ are also set to zero. After the teaching is completed, the robot switches to autonomous execution to follow the produced trajectory $\mathbf{r}_d$. This is achieved by setting the stiffness $\mathbf{K}$ to a high value in order to achieve tracking accuracy. Using the progressive automation approach [49], it is also possible to allow a gradual and seamless transition between teaching and autonomous modes by adjusting $\mathbf{K}$ continuously. However, this would require slightly more time for teaching because the trajectory needs to be demonstrated a few times.

The output of the admittance controller is transformed into joint positions using differential kinematics [50]. These are sent to the robot and are realized by its low-level position controller. Due to an ill-conditioned Jacobian matrix close to singularities, performance constraints [51] are used in both teaching and autonomous mode to prevent reaching such configurations.

Figure 9.3 Teaching by demonstration. The user guides the robot to pick an object from the conveyor and place it at a corner of the container.

9.3.2 Teaching by demonstration

During the teaching phase, the user kinesthetically demonstrates to the robot how to pick an object from the conveyor and how to place it inside the container (Figure 9.3). The purpose of this process is to record the trajectory of the robot and indicate where to grasp the object. The grasp pose is described by the relative pose of the object with respect to the robot's end-effector. To obtain this relative pose, the perception system must first estimate the pose of the object before teaching begins. If there are multiple objects on the conveyor, the pose of all objects must be estimated because it is not known beforehand which object the user will pick for demonstration.

The robot is initially in hand-guidance mode so that the user can guide it over an object on the static conveyor. When the robot is above the target object and ready to grasp it, the user presses a button on the wrist. When the button is pressed, we determine which object is closest to the suction cup of the end-effector based on the estimated poses and store the relative pose of that object. Then the suction cup is activated to grasp the object, and the trajectory recording is started. If no object is detected or the pick fails (based on the pressure feedback from the suction cup), an error message is displayed, and the recording does not start. After the object is successfully picked up, the user guides the robot to place the picked-up object in a corner of the empty container. When the object is in place, the user presses the button again, releasing the object and completing the trajectory recording.

At the end of the trajectory recording, the relative pose of the placed object with respect to the pose of the container is calculated to create the packing plan. The position of the container is determined using markers placed at known locations on it. A single demonstration starting with the pick and ending at the place position of the object is sufficient since in this work we are considering the packing of a single type of product. The motion generation system can then generate the trajectory of the robot from the container toward a new pick position that is provided by the object perception system, using the reversible execution of the DMP.

9.3.3 Dynamic movement primitives

The trajectory that is demonstrated during kinesthetic teaching is encoded in the task space using the DMP formulation proposed in Reference 9 that supports reversibility. Exploiting reversibility enables the use of a single DMP model for placing the object in the container, using forward execution, and picking another object from the conveyor with reverse execution. Moreover, the generalization proposed in

Reference 52 is used, which extends the work from Reference 9 and overcomes the shortcoming of the classical DMP generalization. The new generalization enables efficient scaling in moving targets (picking an object from the conveyor belt) and enforcing via points (for the successful placement of objects inside the container, as described in section 9.3.4) while ensuring a smooth trajectory generation.

The DMP model is governed by the transformation system, which generates the desired reference trajectory, and the canonical system, which controls the evolution of the trajectory. Denoting $\mathbf{y}_d = \mathbf{p}_d$ as the desired Cartesian position, the transformation system's reference trajectory for each time i is given by a sum of L Gaussian kernels weighted by $\mathbf{W}_i$:

$$\mathbf{y}_d(s_i) \triangleq \mathbf{W}_i^T \boldsymbol{\phi}(s_i) \tag{9.2}$$

$$\dot{\mathbf{y}}_d(s_i) = \mathbf{W}_i^T \frac{\partial \boldsymbol{\phi}(s_i)}{\partial s} \dot{s}_i \tag{9.3}$$

$$\ddot{\mathbf{y}}_d(s_i) = \mathbf{W}_i^T \left(\frac{\partial^2 \boldsymbol{\phi}(s_i)}{\partial^2 s} \dot{s}_i^2 + \frac{\partial \boldsymbol{\phi}(s_i)}{\partial s} \ddot{s}_i \right) \tag{9.4}$$

where the weights $\mathbf{W}_i \in \mathbb{R}^{L \times 3}$ are updated online so that the demonstrated acceleration profile is preserved, and the initial, target, via point, and continuity constraints are satisfied as detailed below. The Gaussian kernels are given by:

$$\boldsymbol{\phi}(s)^T = [\psi_1(s) \ \cdots \ \psi_N(s)] / \sum_{k=1}^{L} \psi_k(s), \quad \psi_k(s) = \exp(-h_k(s - c_k)^2)$$

with c_k being the centers and h_k being the inverse widths. The centers c_k are equally spaced in [0, 1] and $h_k = 1/(1.5(c_{k+1} - c_k))^2$. Equations (9.2)–(9.4) are also used for producing the desired orientation trajectory by $\mathbf{y}_d = \log(\mathbf{Q}_d)$. For the orientation DMP, the desired rotational velocity $\boldsymbol{\omega}_d$ and acceleration $\dot{\boldsymbol{\omega}}_d$ can be obtained analytically from $\dot{\mathbf{y}}_d$, $\ddot{\mathbf{y}}_d$ as detailed in Reference 52.

A common canonical system is used to synchronize the position and orientation DMP, given by

$$\ddot{s} = d(\dot{s}_d - \dot{s}), \quad s(0) = s_0, \quad \dot{s}(0) = \dot{s}_d \tag{9.5}$$

where $d > 0$ controls the canonical system's convergence rate to the desired motion duration. Forward execution (from an initial position $\mathbf{y}_0$ to a desired target position $\mathbf{g}$) is realized by setting $s_0 = 0$ and $\dot{s}_d = 1/T_f$, where T_f is the desired motion duration. For reverse execution (i.e., from $\mathbf{g}$ to $\mathbf{y}_0$), we set $s_0 = 1$ and $\dot{s}_d = -1/T_f$.

To achieve the pick of objects from the conveyor and the placement in the container, the DMP weights $\mathbf{W}$ are adapted online by optimizing the demonstrated acceleration profile, while satisfying initial, final, via point, and continuity constraints as proposed in Reference 52. Specifically, in the forward execution, the initial position of the DMP is the position where the robot picked the moving object. The target is the placement position given by the packing plan. Moreover, during the forward execution for the placement, via points are specified to ensure the successful insertion of the object in the container.

The generalization that takes into account the above is achieved by calculating the $\mathbf{W}_i$ at each time-step i so that the following problem is solved [52]:

$$\min_{\mathbf{W}} \sum_{j=1}^{m} \left\| \frac{\partial^2 \hat{\mathbf{y}}_d(s_j)}{\partial^2 s} - \mathbf{W}^T \frac{\partial^2 \boldsymbol{\phi}(s_j)}{\partial^2 s} \right\|_2^2 \tag{9.6a}$$

$$\text{s.t. } \mathbf{W}^T \mathbf{A}(0) = \mathbf{Y}_0 \tag{9.6b}$$

$$\mathbf{W}^T \mathbf{A}(1) = \mathbf{G}_i \tag{9.6c}$$

$$\mathbf{W}^T \mathbf{\Phi}_{v,i} = \mathbf{Y}_{v,i} \tag{9.6d}$$

$$\mathbf{W}^T \mathbf{B}_j = \mathbf{Y}_j, \quad j = 1 \ldots i \tag{9.6e}$$

where $\frac{\partial^2 \hat{\mathbf{y}}_d(s_j)}{\partial^2 s} = \mathbf{W}_0^T \frac{\partial^2 \boldsymbol{\phi}(s_j)}{\partial^2 s}$ is the learned acceleration profile, with

$$\mathbf{W}_0 = \operatorname{argmin}_{\mathbf{W}} \sum_{j=1}^{m} \|\mathbf{y}(s_j) - \mathbf{W}^T \boldsymbol{\phi}(s_j)\|_2^2$$

where $\mathbf{y}$ is the demonstrated trajectory, and $\hat{\mathbf{y}}_d(s) = W_0^T \boldsymbol{\phi}(s)$ is the learned trajectory. The matrices in the first two equality constraints, (9.6b), (9.6c), with $\mathbf{A}(s) = [\boldsymbol{\phi}(s)\ \dot{\boldsymbol{\phi}}(s)\ \ddot{\boldsymbol{\phi}}(s)]$, enforce the initial state $\mathbf{Y}_0 = [\mathbf{y}_0\ \mathbf{0}_{n\times 1}\ \mathbf{0}_{n\times 1}]$ and final state $\mathbf{G}_i = [\mathbf{g}_i\ \mathbf{0}_{n\times 1}\ \mathbf{0}_{n\times 1}]$ conditions. Via point constraints are defined in (9.6d), where $\mathbf{Y}_{v,i} = [\mathbf{y}_{v,1} \ldots \mathbf{y}_{v,L_i}]$ contains the via points, with L_i their number at the current time-step i, and $\mathbf{\Phi}_{v,i} = [\boldsymbol{\phi}(s_{v,1}) \ldots \boldsymbol{\phi}(s_{v,L_i})]$. The constraints in (9.6e) with $\mathbf{B}_j = \mathbf{B}(s_j) = [\boldsymbol{\phi}(s_j)\ \dot{\boldsymbol{\phi}}(s_j)\ \ddot{\boldsymbol{\phi}}(s_{j-1})]$ and $\mathbf{Y}_j = [\mathbf{y}(s_j)\ \dot{\mathbf{y}}(s_j)\ \ddot{\mathbf{y}}(s_{j-1})]$ encode the current and all previous state constraints up to time-step i to ensure a smooth trajectory generation when via points are introduced or when the target pose of the object to be picked from the conveyor is updated. The solution to problem (9.6) is obtained by recursively updating the DMP weights $\mathbf{W}_i$ at each time-step i as detailed in Reference 52.

9.3.4 Packing plan and via points

Determining the placement pose of each object within the container is equivalent to the pallet loading problem [53] that focuses on space optimization to pack the maximum quantity of goods onto the pallet. To narrow down the search space, it is assumed that all objects are of the same known orthogonal size, and the container dimensions are given, which is a valid assumption for most manufacturing lines.

The proposed packing planner is a computationally lightweight method to determine the placement locations based on the first object that is placed by the operator in the initial demonstration. As shown in Figure 9.4, the method is based on a depth-first search algorithm and places objects in parallel to each other as long as possible before backtracking to another row. The search tree considers for 90° rotation of objects around the vertical axis. The algorithm works as follows:

1. After the operator has demonstrated the placement of the first object at a corner of an empty container, the relative pose of the object with respect to the container is calculated. From this pose, the object alignment to the container walls is checked. If it is below 5°, the gripper releases the object to that position when the operator presses the button. If not, the object is not released and an appropriate message is indicated to correct the alignment. The 5° threshold depends on the application and the allowed clearance between objects.

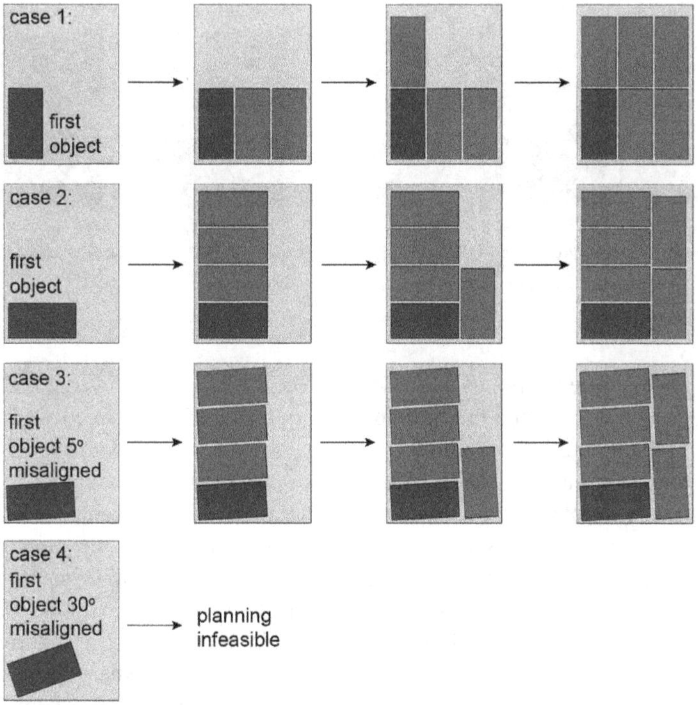

Figure 9.4 Packing plan in four different use-cases

2. After the object is released, the algorithm calculates the next placement position to be side-by-side and in parallel to the previous object. This step is repeated until the empty space can be filled with this orientation.
3. The algorithm searches for filling the remaining empty space by rotating the object by 90°.
4. When the layer is filled, the same process is repeated for as many layers can fit in the container. Unlike pallets where each layer needs to be stacked differently for maintaining the stability of the structure, a container does not have this requirement because the objects are constrained by the side walls.

A container such as a carton box presents an additional challenge compared to a pallet because of the side walls that need to be considered as obstacles during motion planning. This means that the clearance from both the side walls and the previously placed objects needs to be taken into account when scaling the DMP trajectory. Depending on the obstacles that surround the target placement position in the container, via points are calculated automatically and are imposed on the generated placement trajectory. The via points are calculated so that they apply offsets in the position and orientation trajectory close to the placement target. Three cases are distinguished for all possible placement positions, as shown in Figure 9.5. For each of these three cases, we apply different types of via points that are shown in Figure 9.6.

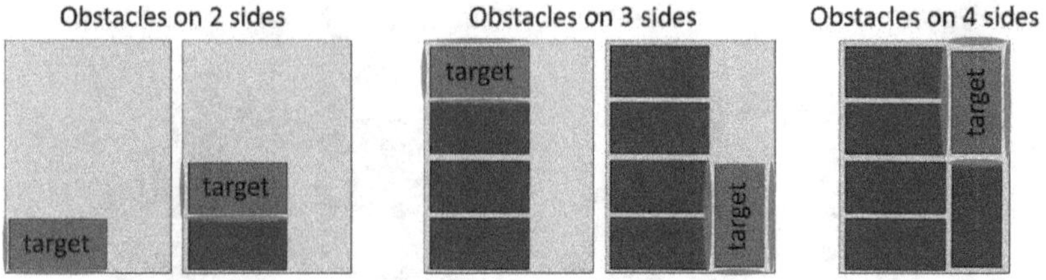

Figure 9.5 The three cases of adjacent obstacles (either container wall or other objects) at the target position of an object

We generate the via points as follows. Given the height of the container d_0^z, the height of the object d_0^Z, and the desired vertical distance d_{vp} between via point, we determine the number of via points $n_{vp} = (d_c^z - 0.5d_0^z)/d_{vp}$ rounded to the nearest integer. The top via point will then be placed at the same height as the container and the bottom one at a distance $0.5d_o^z$ from the container floor. The horizontal offset and the rotation depend on the surrounded obstacles. In the simplest case, the target position is surrounded by obstacles on two sides (Figure 9.6a). Because of the empty space in the remaining two sides, we apply a gradual translational offset with the maximum value equal to a quarter of the object's dimension in that axis. The orientation of the via points in this case is equal to the target one. In the case where three sides are surrounded (Figure 9.6b,c), a translational offset is applied toward the empty space and a rotation of $\pi/6$ rad around that direction (−X in Figure 9.6b). The placement of the last object in each layer is equivalent to a peg-in-hole problem. In this case, two sequential rotations are applied to facilitate insertion (−X and −Y; Figure 9.6c), inspired by Reference 54.

9.4 Object perception

To achieve the objective of versatile robot packing, it is necessary to obtain the position and orientation of target objects (6D pose) while they are moving on the conveyor that feeds them to the robot workspace. Relying on an accurate perception system, the need for mechanical guides and hard automation is lifted, increasing flexibility and reducing associated costs. We propose a pipeline that consists of three stages: instance segmentation, pose estimation, pose refinement, and tracking. In the first stage, the RGB image from the camera is passed through a Mask R-CNN model to detect whether there are objects in the scene and to determine their corresponding regions on the image. In the second stage, the cropped region (mask) for each of the detected instances is used to determine the 6D pose fusing the depth information as well. The third stage involves the refinement of the detected 6D pose to improve accuracy [27, 30] and includes the tracking of the object even when it is occluded by the robot. The complete perception pipeline for pose estimation is shown in Figure 9.7.

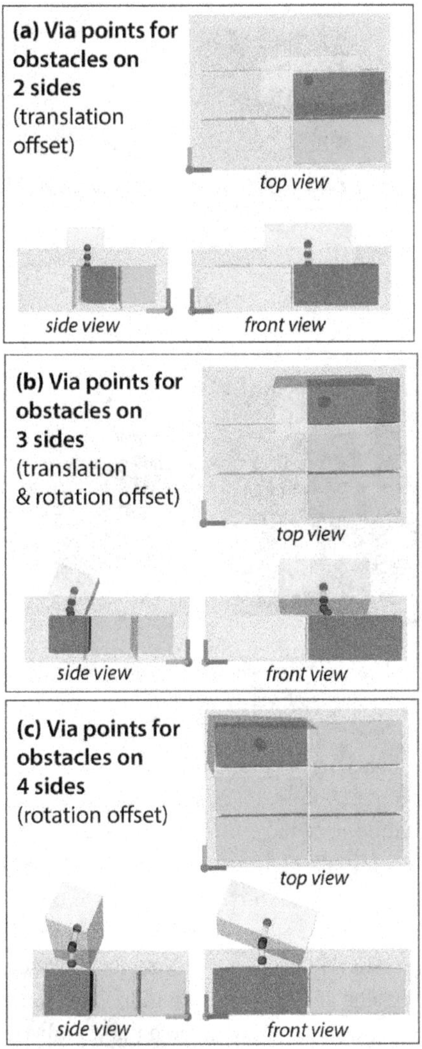

Figure 9.6 Via points for translation or rotation offset in the placement trajectory. The position of each via point is shown as a dark red sphere. Solid green is the objects already placed (occupied space), and dark blue is the target pose for the object being placed, which is shown in light blue at the top via point. Light yellow is the next objects in the plan (free space).

9.4.1 Instance segmentation stage

The first stage in the pipeline is instance segmentation [21]. We use a Mask R-CNN model [55] with a pretrained ResNet-50 convolutional backbone [16]. The network is

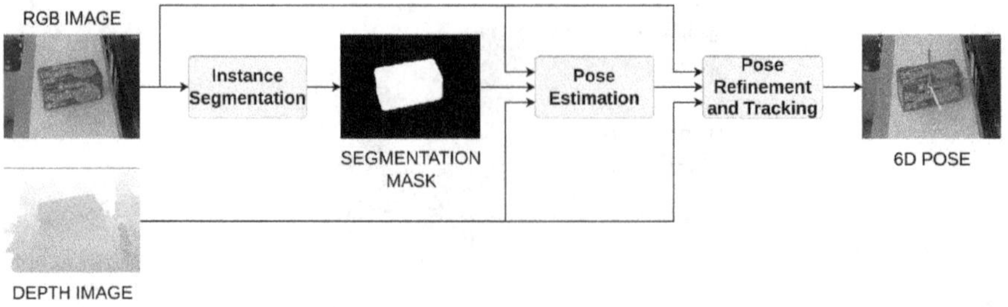

Figure 9.7 Perception pipeline for pose estimation

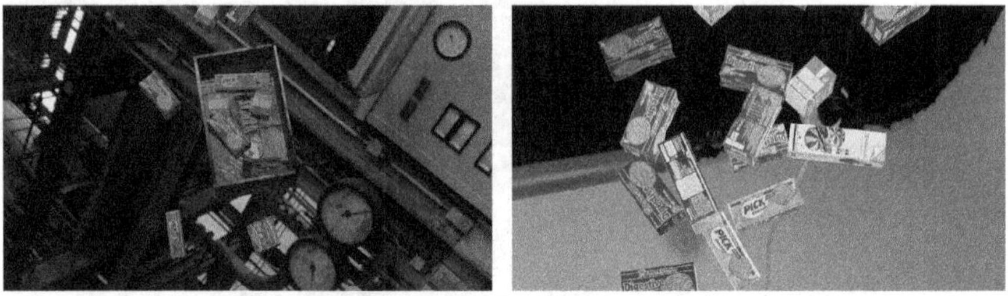

Figure 9.8 Synthetic images rendered in Blender for the Mask R-CNN training

trained using a synthetic dataset generated in Blender† using the 3D model of the object of interest (shape and texture). Additional object models are placed in the Blender scene in order to create a cluttered and, thus, more realistic scene, as shown in Figure 9.8. Each scene is created by initializing multiple objects with random poses and dropping them onto a surface or inside a box. The physics engine included in Blender is used to simulate the interactions between the objects before rendering. Panoramic high dynamic range imaging (HDRI) images were used for the background of each render since they include both color and light information. The light information from the HDRI image is used to illuminate the objects in a photorealistic way. For each render, the background is sampled at random from a selection of 30 HDRI images. A total of 6,000 images are rendered in order to train, evaluate, and tune the instance segmentation network. This stage uses the RGB image as input and outputs the segmentation mask for each detected instance.

9.4.2 Pose estimation stage

For the pose estimation of the objects, a feature-matching algorithm is used. Initially, we use the synthetic images of the object that were generated in the previous step

†https://www.blender.org/, https://polyhaven.com/hdris

along with their pose information to create a reference dictionary using scale-invariant feature transform (SIFT) features. SIFT features are local descriptors invariant to the scale, rotation, and translation and robust to illumination changes and noise in the image domain. The SIFT features of the dictionary are mapped to 3D points of the object surface.

During the estimation stage, the masks from the instance segmentation are used as input in order to extract only the region from the color and depth images that contain the corresponding object. The SIFT features of the extracted region are then calculated and mapped to 2D pixels on the image, creating a query dictionary. The SIFT descriptors of the reference and query dictionaries are then matched using the nearest neighbor search. For the nearest neighbor search, a fast approximate algorithm was used from the Fast Library for Approximate Nearest Neighbors. The resulting matched pairs of the 3D object points from the reference dictionary and the 2D pixels from the query dictionary are then used to solve the Perspective-n-Point pose computation problem. A Random sample consensus (RANSAC) based method is used to account for erroneous matches. RANSAC is an iterative method for model estimation in the presence of outliers. During each iteration, a random subset of the matches is selected, and the pose is calculated. After the prescribed number of iterations, the pose with the most inliers is selected.

This process for pose estimation is repeated for each object that was detected in the previous step.

9.4.3 Pose refinement and tracking stage

Although RGB-D cameras can provide color and depth images at a high frame rate (30–60 fps), the computation required for pose estimation is quite heavy. Despite the fact that it is being performed on a high-end GPU with CUDA capabilities, it takes at least 100 ms so that the maximum refresh rate of the object's pose is at 10 Hz. This delay causes problems in the case of moving objects since the calculated pose will not match the actual one.

In this stage, we use the pose estimation from the previous step and perform an iterative pose refinement at a higher rate to accurately track the moving object. Specifically, we use the ICG pose refiner and tracker [27] because of its state-of-the-art performance with respect to calculation speed. ICG is a probabilistic tracker that fuses color and depth information and only requires object geometry. It deploys correspondence lines and points between consecutive image frames to iteratively refine the 6D pose of the object. Since ICG is based on a local optimization technique, it requires an initial estimate of the pose that is given from the output of the pose estimator described in section 9.4.2. The result is reliable and accurate real-time 6D pose tracking of the target object.

9.4.4 Occlusion detection and handling

The perception system uses an RGB-D camera with the robot either fully or partially in its field of view. While picking up objects, the robot may either partially or completely occlude the moving object, which may affect the detection accuracy of the perception pipeline. An error in perception can seriously affect the robot's performance, preventing it from successfully completing its task and posing a risk of damage due to possible misbehavior. This can happen because the motion generation module of section 9.3.3 constantly adapts to changes in the target pose during picking according to the estimation of the object perception subsystem.

With the URDF model and the transformation of the camera's coordinate frame from the robot base, we can use the joint position measurements of the robot to render the VDI. The VDI is a noise-free depth image of the robot as seen by the camera, from which a perception system can infer whether the target is partially or fully occluded, improving the robustness of the overall perception system.

To render the VDI, we use a virtual camera with a pinhole camera model, where the intrinsic and extrinsic parameters correspond to the known parameters of the actual camera. The 3D model is taken from the URDF and the joint positions from the actual robot. For the implementation of the renderer, we use the OpenGL library. The frame rate of the virtual camera matches that of the real camera. The renderer first obtains the joint positions of the robot, then transforms the pose of each robot link from the world frame to the camera frame, and finally renders all 3D models to generate each image. The output is a noise-free VDI of the robot, as shown in Figure 9.9.

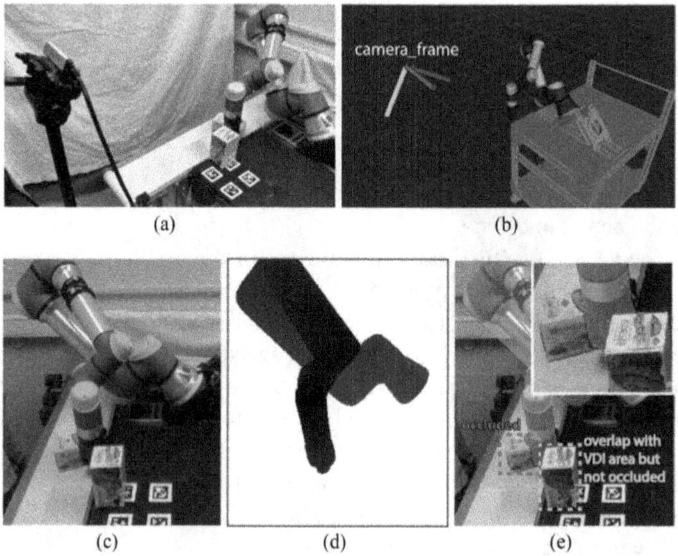

Figure 9.9 (a) Setup with a robot, an RGB-D camera, and two objects (one in front and one behind the robot), (b) the digital visualization, (c) the color view from the camera, (d) the rendered VDI including the end-effector, and (e) the augmented view highlighting the areas occluded by the robot

Using the VDI, we can determine when the object becomes partially or fully occluded and filter the object's pose that is derived by the object perception pipeline. If an occlusion is detected, it is handled by predicting the poses of the object based on the current non-occluded estimates of pose and velocity as in Reference 56. The velocity is estimated by using a least squares regression on a window of the last 20 pose estimates. These poses are then provided as targets for the motion generation module. When the object is no longer occluded, we switch back to the perception pipeline.

9.5 Human perception

The safety of people working near the robot is an important consideration in any collaborative application. The current technical specification ISO/TS 15066 defines the mode *speed and separation monitoring* to ensure the safety of people in the vicinity when the robot is working autonomously. In this mode, a separation distance S_p between humans and the robot must be maintained for the latter to operate autonomously. This distance is defined as the minimum distance from any point of the robot to any point of a human in the robot's work cell. The distance depends on the maximum operating speed of the robot, the reaction time of the perception system, and the robot's stopping distance so that it can safely come to a stop before coming into contact with the human. If the current minimum distance d_h is below the safety distance S_p, the robot must stop and not resume operation until $d_h > S_p$. In this approach, a safety zone is defined that allows execution at full speed only when humans are outside this zone and forces the robot to stop when $d_h < S_p$.

In this work, we apply two zones by introducing the distance S_r with $S_r > S_p$ that allows autonomous execution at a reduced speed when $S_p < d_h < S_r$. By scaling down the velocity of the robot in this area, we can relax S_p by reducing the minimum separation distance and allow safe operation of the robot even when the human is within the robot's workspace. The reduced speed zone and the safety zone are illustrated in Figure 9.10.

For human detection, an overhead wide-angle RGB-D camera is used to cover the robot's workspace. The AlphaPose [48] library is used for 2D human skeleton detection on the color image. The human skeleton is given as key-point pixels on the color image, which represent the main joints of the human body, including arms, legs, torso, and head. The 2D key-points are deprojected using the depth information into 3D points with respect to the robot frame in order to calculate the minimum distance from the robot. To avoid erroneous depth information due to the occlusion of a human key-point from the robot, the VDI described in section 9.4.4 is also used here to detect occlusions (Figure 9.11). Then, the minimum distance d_h between the robot and the key-points on human skeleton is calculated as in Reference 44.

Finally, the speed of the generated DMP trajectory is scaled with phase stopping by scaling in the canonical system of (9.5), as follows:

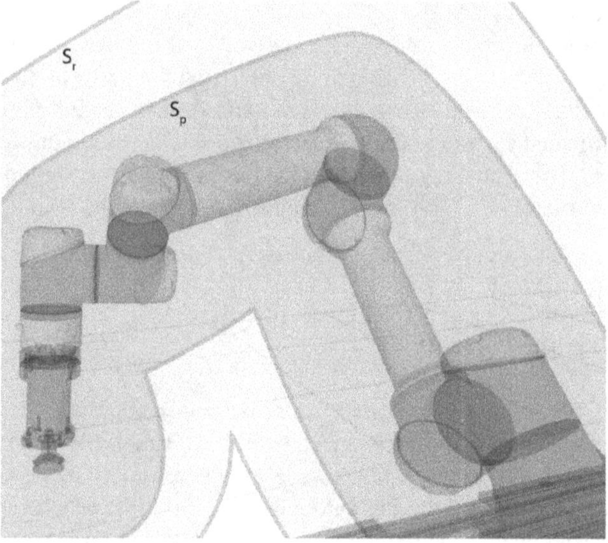

Figure 9.10 *Safety zones of the robot arm for human safety (approximate visualization)*

$$\dot{s} = \begin{cases} 1/T_f, & \text{if } d_h \geq S_r \\ \frac{1}{T_f}\left(\frac{d_h - S_p}{S_r - S_p}\right)^3, & \text{if } S_p \leq d_h < S_r \\ 0 & \text{if } d_h < S_p \end{cases} \tag{9.7}$$

When the human is in the reduced speed zone, (9.7) slows down the time evolution of the reference trajectory and hence of the robot speed. A nonlinear function of the minimum distance d_h is used to allow prompt reduction of robot velocity when the

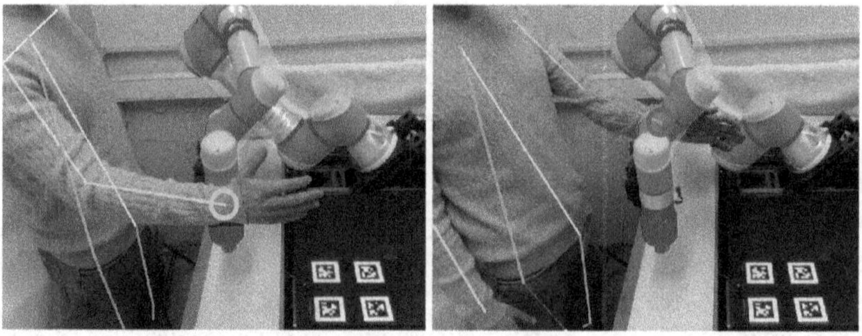

Figure 9.11 *Human skeleton is detected by AlphaPose, and the closest skeleton joint to the robot is determined (circles). The VDI is also used to detect occluded parts of the human skeleton.*

human approaches the safety zone. Within the safety zone, the DMP is paused with $\dot{s} = 0$, and the velocity of the robot becomes zero.

9.6 Experimental validation

To evaluate the proposed packing system, we perform a dense packing task using the setup of Figure 9.1. It consists of a UR5e 6-DOF robotic arm with an EPick vacuum gripper. The robot has an embedded force/torque sensor on the wrist that is used to obtain the user-applied wrench for the admittance controller. A conveyor belt feeds objects to the robot workspace to be packed into a carton box. During autonomous execution, the conveyor stops automatically when objects reach the end of it, using a proximity sensor. A RealSense D415 camera is used for the object perception subsystem and a ZED2 for human perception. The former has better accuracy due to its narrow field of view, but the latter has a wide-angle lens that can cover a large workspace of the robot during human detection. The proposed architecture is implemented using ROS2 on a Ryzen 3700x CPU and a GTX1060 GPU.

Both cameras capture color and depth images at 30 fps. In the object perception pipeline, the pose estimation stage has the slowest performance and can provide estimates at a rate of 10 Hz. However, pose refinement occurs at 30 Hz. The refined pose estimates are then used to obtain accurate velocity estimates for the tracking stage. The human perception subsystem provides the minimum distance of the robot from nearby humans at a rate of 15 Hz. The VDI of the robot arm is rendered twice at 30 fps, once from the viewpoint of the object detection camera and once from the viewpoint of the human detection camera.

9.6.1 Teaching by demonstration and pack planning

A user first teaches the robot by demonstration, as shown in Figure 9.12. The parameters used for the inertia **M** of the admittance controller are 5 kg for the translational components and 0.5 kgm^2 for the rotational components. For the damping **M**, they are 80 Ns/m and 0.5 Nms/rad, respectively. The stiffness term **K** is zero. Before the teaching starts, the perception pipeline has estimated the pose of all visible objects on the static conveyor. The user guides the robot to one of these objects and presses

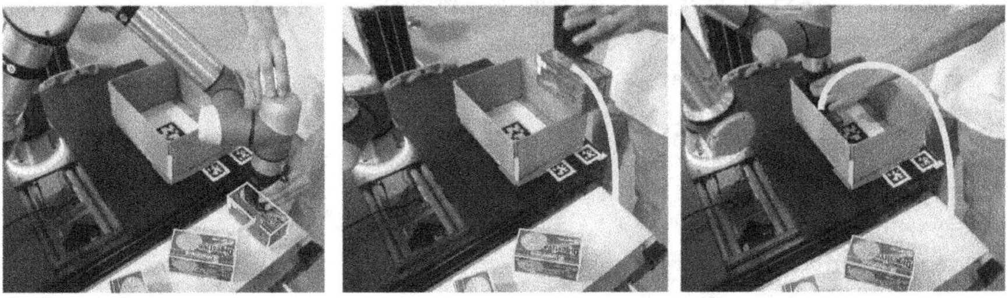

Figure 9.12 The user teaches the robot by demonstration, picking an object from the static conveyor and placing it at a corner of the container

Figure 9.13 Automatic generation of packing plan after the teaching is completed

the button, and the recording of the robot's trajectory begins. At the same time, the relative grasping pose of the object is calculated. Then the user places the picked object to a corner of the container and presses the button to release it. After the object alignment is checked, the packing plan is automatically generated in real time, as shown in Figure 9.13. The duration of the pick-and-place demonstration is 9 seconds, which shows how fast and easy the teaching stage is. After teaching, the conveyor starts moving at a constant speed and feeds objects to the robot for autonomous packing.

9.6.2 Detection of moving objects with partial occlusions

During autonomous task execution, the robot may partially or fully occlude the moving object, leading to wrong estimates of its pose. Using the VDI, the occlusion percentage of the target object is calculated. If the occlusion percentage is more than 10%, the moving object's pose is calculated using the last non-occluded pose and velocity estimates. If the percentage falls below this threshold, the tracker again uses the estimates from the object detection pipeline.

Figure 9.14 shows how the proposed perception subsystem detects a moving object with partial occlusions under various orientations and occlusions. When the object is partially occluded by the robot, the object detection system provides inaccurate estimates of the object's pose. Nevertheless, tracking with the VDI provides a much more accurate estimate of the object pose, increasing the robustness of the overall packing system.

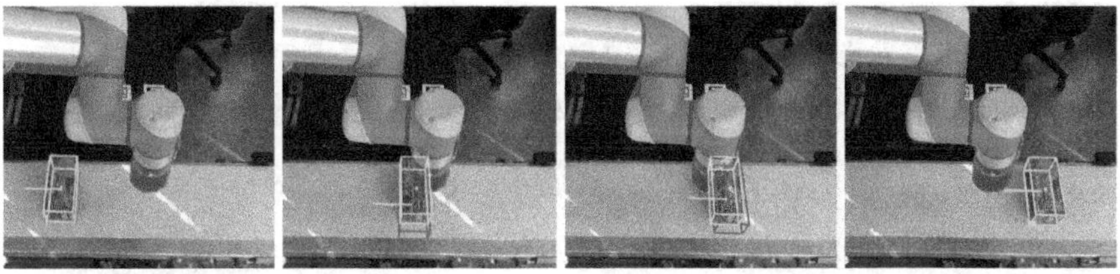

Figure 9.14 Refined tracking of objects moving on the conveyor. When the object is partially occluded by the robot, the object detection systems provide inaccurate estimates of the object's pose (blue bounding boxes). Utilizing the VDI, we can accurately track the object (green bounding boxes) based on its last non-occluded position and velocity estimates.

9.6.3 Autonomous execution with via points

In the autonomous execution, the stiffness term **K** is set to 5000 N/m and 300 Nm/rad. The damping **C** is also increased to 350 Nm/s and 10 Nms/rad, respectively, to avoid oscillations in the behavior of the admittance. Once the perception pipeline detects objects moving on the conveyor, the pose of the object closer to the end of the conveyor is specified as the target for the DMP to generate the trajectory and pick it. The DMP formulation makes it possible to reach the object precisely either it is moving (Figure 9.15a) or is stopped at the end of the conveyor (Figure 9.15b).

(a)

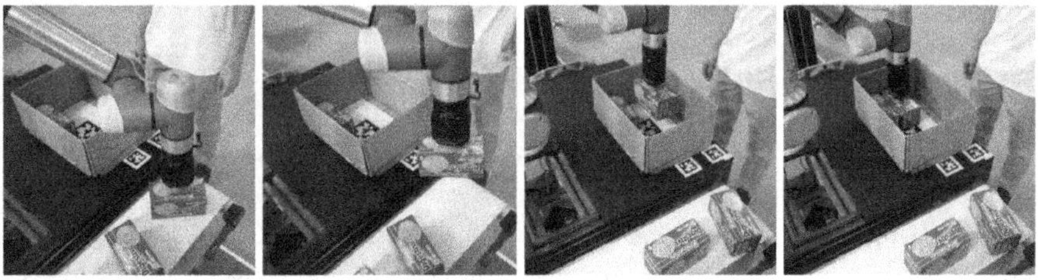

(b)

Figure 9.15 Autonomous execution of the pick and place by the robot. (a) Picking a moving object. (b) Picking an object that has stopped at the end of the conveyor and placing it at the container.

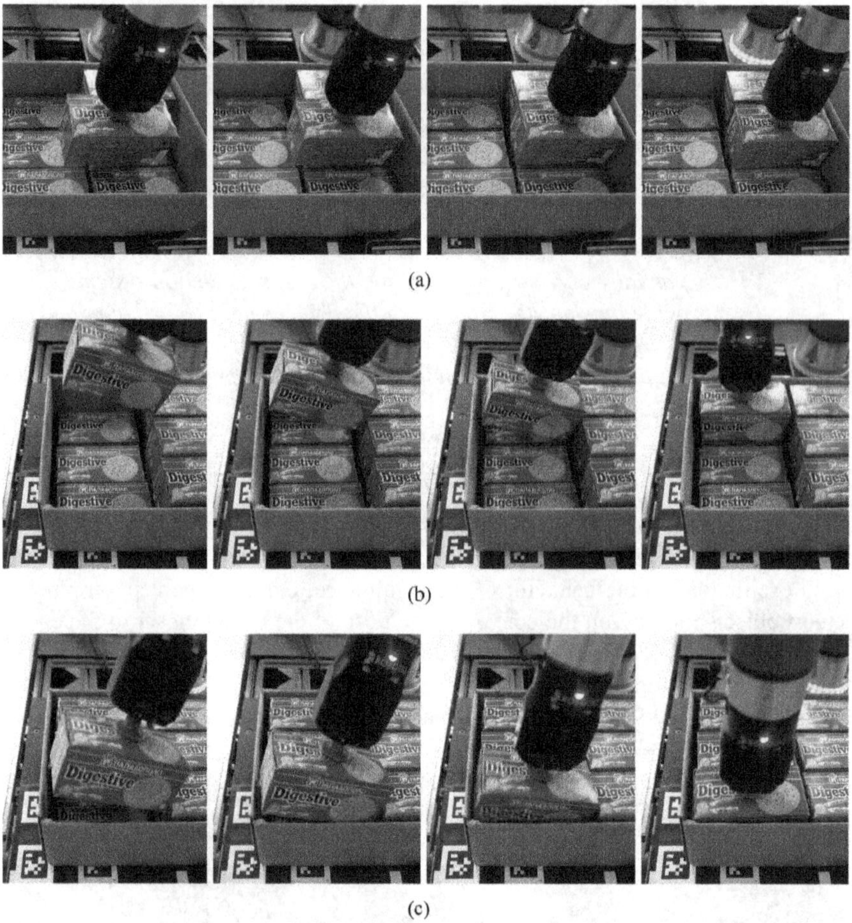

Figure 9.16 Placement offset applied by via points for all possible cases. (a) Obstacles on two sides, (b) obstacles on three sides, and (c) obstacles on all four sides.

After pickup, the target for placement is provided by the packing plan, including the via points, depending on the location within the container. The position and orientation offset due to the via points are shown for all possible cases in Figure 9.16.

9.6.4 Human safety

To demonstrate the safety aspect of the proposed system, we perform an experiment where a human enters the workspace of the robot during the autonomous execution, as shown in Figure 9.17a. The distance S_p that defines the safety zone is selected equal to 0.5 m, taking into consideration the reaction time of the human perception subsystem and a safety margin because of the skeleton tracking. This is required

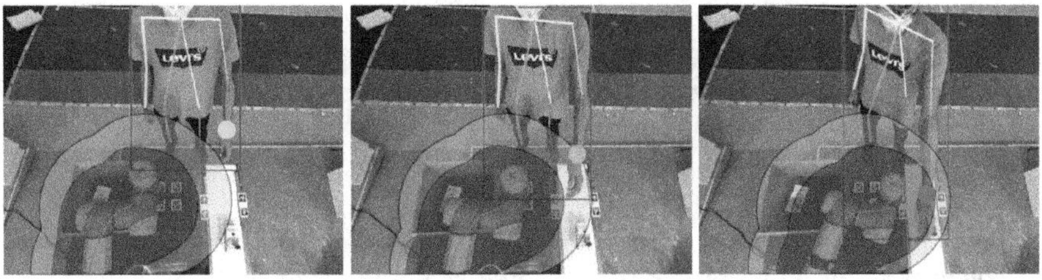

(a)

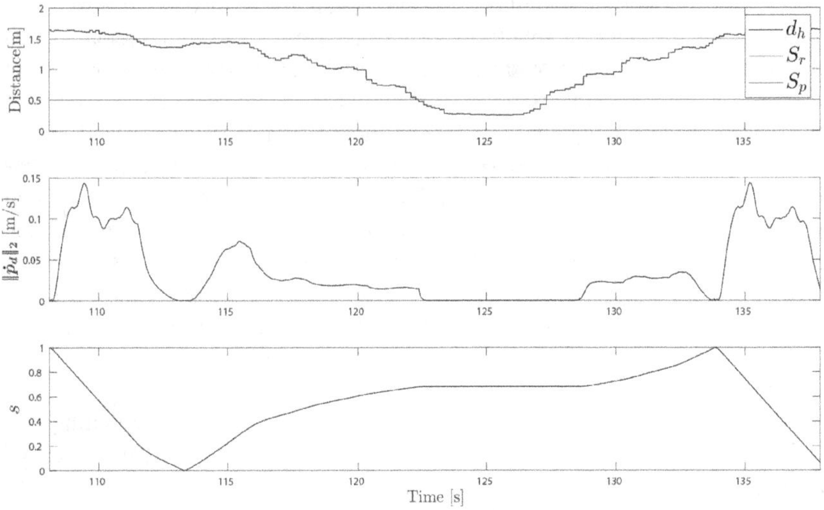

(b)

Figure 9.17 When the human enters the reduced speed area, robot velocity is scaled down according to the minimum distance. When the safety zone is crossed the robot stops. (a) The view from the human detection camera. (b) Plot of the minimum distance, the velocity of the robot, and the phase variable.

because the skeleton algorithm provides a simplified representation of the human body. In this experiment, for example, the skeleton algorithm detects that the wrist of the human is closer to the robot without considering the human's fingers. Although this is sufficient for such applications, skeleton tracking algorithms already support finger tracking and body shape estimation if it is required. The distance S_r that defines the reduced speed zone is set equal to 1.5 m.

The diagram of Figure 9.17b illustrates the minimum distance d_h, the norm of the velocity $\dot{\mathbf{p}}_d$ that is produced by the DMP, and the phase variable that is scaled according to the minimum distance. Initially, the human is far from the robot which operates autonomously with a maximum velocity of 0.15 m/s, and the phase variable evolution is linear. Then the human brings his arm close to the robot, entering the reduced speed zone. The closer the human gets to the safety zone, the velocity of the robot is scaled down. When the human enters the safety zone, the generated velocity of the robot quickly drops to zero, bringing the robot to a stop. After the human leaves the safety zone, the robot resumes at a reduced speed.

9.7 Conclusion

In conclusion, our work proposes a versatile and efficient solution for automating dense packing tasks in an industrial environment using teaching by demonstration and machine vision. By combining these two techniques, we enable the user to easily and quickly program the robot's pick-and-place motion pattern without any coding experience while also providing the robot with the ability to recognize its environment, adapt its movements, and remain safe for human interaction. We propose a novel machine vision pipeline that has high accuracy and can handle partial or full occlusions of objects by the robot body, which allows the system to accurately detect and track objects moving on a conveyor. Furthermore, the system can detect humans within the robot's workspace and adjust the robot's speed to maintain the minimum distance between the robot arm and the human body. Our experiments validate the effectiveness of the system in a series of dense packing tasks. Ultimately, our goal is to provide a portable and flexible work cell with a robotic arm that can be easily taught by personnel to pack products on a production line, improving efficiency and reducing costs.

While our proposed solution shows promising results for automating dense packing tasks in an industrial environment, there are some limitations to consider. One major limitation is that the system's performance may be affected by changes in lighting conditions, as the machine vision pipeline heavily relies on visual data. Additionally, the performance of the perception pipeline relies on the quality and quantity of the training data. Gathering sufficient and diverse training data can be challenging and time-consuming, especially for applications with a large variety of objects. Even with sufficient training data, the perception pipeline may struggle to generalize to new objects that were not included in the training set. This is particularly relevant for applications with a high variety of objects or where new objects are regularly introduced to the system. However, our proposed solution has potential applications in a wide range of industrial settings beyond packing tasks. For example, it could be adapted to sort and assemble products in a warehouse or to pick-and-place items in a manufacturing facility. The teaching by demonstration approach also makes the system accessible to a broader range of users, including those without extensive programming knowledge. In the future, we aim to continue improving the system's capabilities and exploring new potential applications.

Acknowledgments

Cofinanced by the European Regional Development Fund of the European Union and Greek national funds through the Operational Program Competitiveness,

Entrepreneurship and Innovation, under the call RESEARCH – CREATE – INNOVATE (project code: T2EDK-02358).

References

[1] Shome R., Tang W.N., Song C. 'Towards robust product packing with a minimalistic end-effector'. *2019 International Conference on Robotics and Automation (ICRA)*; Montreal, QC, Canada, IEEE, 2019. pp. 9007–13.
[2] Sucan I.A., Chitta S. MoveIt [online]. Available from moveit.ros.org
[3] Wang F., Hauser K. 'Dense robotic packing of irregular and novel 3D objects'. *IEEE Transactions on Robotics*. 2021, vol. 38(2), pp. 1160–73.
[4] Wang F., Hauser K. 'Robot packing with known items and nondeterministic arrival order'. *IEEE Transactions on Automation Science and Engineering*. 2020, vol. 18(4), pp. 1901–15.
[5] Yang Z., Yang S., Song S. 'Packerbot: variable-sized product packing with heuristic deep reinforcement learning'. *2021 IEEE/RSJ International Conference on Intelligent Robots and Systems (IROS)*; Prague, Czech Republic, IEEE, 2021. pp. 5002–08.
[6] Hong Y.D., Kim Y.J., Lee K.B. 'Smart pack: online autonomous object-packing system using RGB-D sensor data'. *Sensors (Basel, Switzerland)*. 2020, vol. 20(16), 4448.
[7] Billard A.G., Calinon S., Dillmann R. 'Learning from humans' in *Springer Handbook of Robotics*. Springer International Publishing; 2016. pp. 1995–2014.
[8] Ijspeert A.J., Nakanishi J., Hoffmann H., Pastor P., Schaal S. 'Dynamical movement primitives: learning attractor models for motor behaviors'. *Neural Computation*. 2013, vol. 25(2), pp. 328–73.
[9] Sidiropoulos A., Doulgeri Z. 'A reversible dynamic movement primitive formulation'. *2021 IEEE International Conference on Robotics and Automation (ICRA)*; IEEE, 2021. pp. 3147–53.
[10] Sidiropoulos A., Papageorgiou D., Doulgeri Z. 'A novel framework for generalizing dynamic movement primitives under kinematic constraints'. *Autonomous Robots*. 2023, vol. 47(1), pp. 37–50.
[11] Gribovskaya E., Billard A. 'Learning nonlinear multi-variate motion dynamics for real-time position and orientation control of robotic manipulators'. *2009 9th IEEE-RAS International Conference on Humanoid Robots*; 2009. pp. 472–77.
[12] Pastor P., Kalakrishnan M., Meier F., *et al.* 'From dynamic movement primitives to associative skill memories'. *Robotics and Autonomous Systems*. 2013, vol. 61(4), pp. 351–61.
[13] Paraschos A., Daniel C., Peters J., *et al.* 'Probabilistic movement primitives' *In: advances in neuralinformation processing systems*; 2013. pp. 2616–24.
[14] Chen J., Zhang L., Liu Y., Xu C. 'Survey on 6D pose estimation of rigid object'. *2020 39th Chinese Control Conference (CCC)*; Shenyang, China, IEEE, 2020. pp. 7440–45.

[15] LeCun Y., Bengio Y. 'Convolutional networks for images, speech, and time series'. *The Handbook of Brain Theory and Neural Networks*. 1995, vol. 3361, p. 10.

[16] He K., Zhang X., Ren S., Sun J. 'Deep residual learning for image recognition'. *Proceedings of the IEEE Conference on Computer Vision and Pattern Recognition*; 2016. pp. 770–78.

[17] Tan M., Le Q. 'Efficientnet: rethinking model scaling for convolutional neural networks'. *In: International Conference on Machine Learning. PMLR*; 2019. pp. 6105–14.

[18] Vaswani A., Shazeer N., Parmar N., *et al*. 'Attention is all you need'. *Advances in Neural Information Processing Systems*. 2017, vol. 30.

[19] Zhu X., Su W., Lu L., *et al*. 'Deformable DETR: deformable transformers for end-to-end object detection'. *arXiv preprint arXiv:201004159*. 2020.

[20] Liu Z., Lin Y., Cao Y., *et al*. 'Swin transformer: hierarchical vision transformer using shifted windows'. *2021 IEEE/CVF International Conference on Computer Vision (ICCV)*; 2021. pp. 10012–22.

[21] Hafiz A.M., Bhat G.M. 'A survey on instance segmentation: state of the art'. *International Journal of Multimedia Information Retrieval*. 2020, vol. 9(3), pp. 171–89.

[22] Wang X., Zhang R., Kong T., *et al*. 'Solov2: dynamic and fast instance segmentation'. *Advances in Neural Information Processing Systems*. 2020, vol. 33, pp. 17721–32.

[23] Girshick R., Donahue J., Darrell T., Malik J. 'Rich feature hierarchies for accurate object detection and semantic segmentation'. *2014 IEEE Conference on Computer Vision and Pattern Recognition (CVPR*); Columbus, OH, USA, 2014. pp. 580–87.

[24] Girshick R. 'Fast r-cnn'. *2015 IEEE International Conference on Computer Vision (ICCV)*; 2015. pp. 1440–48.

[25] He K., Gkioxari G., Dollar P., Girshick R. 'Mask R-CNN'. *2017 IEEE International Conference on Computer Vision (ICCV)*; 2017. pp. 2961–69.

[26] Hinterstoisser S., Holzer S., Cagniart C. 'Multimodal templates for real time detection of texture-less objects in heavily cluttered scenes'. *2011 International Conference on Computer Vision*; IEEE, 2011. pp. 858–65.

[27] Stoiber M., Sundermeyer M., Triebel R. 'Iterative corresponding geometry: fusing region and depth for highly efficient 3D tracking of textureless objects'. *2022 IEEE/CVF Conference on Computer Vision and Pattern Recognition (CVPR)*; 2022. pp. 6855–65.

[28] Peng S., Liu Y., Huang Q., Zhou X., Bao H. 'PVNet: pixel-wise voting network for 6dof pose estimation'. *2019 IEEE/CVF Conference on Computer Vision and Pattern Recognition (CVPR)*; 2019. pp. 4561–70.

[29] Wang C., Xu D., Zhu Y., *et al*. 'Densefusion: 6D object pose estimation by iterative dense fusion'. *2019 IEEE/CVF Conference on Computer Vision and Pattern Recognition (CVPR)*; 2019. pp. 3343–52.

[30] Iwase S., Liu X., Khirodkar R., Yokota R., Kitani K.M. 'Repose: fast 6D object pose refinement via deep texture rendering'. *2021 IEEE/CVF International Conference on Computer Vision (ICCV)*; 2021. pp. 3303–12.

[31] Xu Y., Lin K.Y., Zhang G., Wang X., Li H. 'Rnnpose: Recurrent 6-DoF object pose refinement with robust correspondence field estimation and pose optimization'. *2022 IEEE/CVF Conference on Computer Vision and Pattern Recognition (CVPR)*; 2022. pp. 14880–90.

[32] Lipson L., Teed Z., Goyal A., Deng J. 'Coupled iterative refinement for 6D multi-object pose estimation'. *2022 IEEE/CVF conference on computer vision and pattern recognition*; 2022. pp. 6728–37.

[33] Yoshioka K., Okuni H., Ta T.T., Sai A. 'Through the looking glass: diminishing occlusions in robot vision systems with mirror reflections'. *2021 IEEE/RSJ International Conference on Intelligent Robots and Systems (IROS)*; IEEE, 2021.

[34] Luo Y., Dong K., Zhao L., *et al.* 'Calibration-free monocular vision-based robot manipulations with occlusion awareness'. *IEEE Access*. 2021, vol. 9, pp. 85265–76.

[35] Wüthrich M., Pastor P., Kalakrishnan M., Bohg J., Schaal S. 'Probabilistic object tracking using a range camera'. *2013 IEEE/RSJ International Conference on Intelligent Robots and Systems*; IEEE, 2013. pp. 3195–202.

[36] Isobe Y., Masuyama G., Umeda K. 'Occlusion handling for a target-tracking robot with a stereo camera'. *Robomech Journal*. 2018, vol. 5(1), pp. 1–13.

[37] Shu G., Dehghan A., Oreifej O., *et al.* 'Part-based multiple-person tracking with partial occlusion handling'. *In: 2012 IEEE Conference on Computer Vision and Pattern Recognition*; IEEE, 2012. pp. 1815–21.

[38] Magrini E., Flacco F., De Luca A. 'Estimation of contact forces using a virtual force sensor'. *2014 IEEE/RSJ International Conference on Intelligent Robots and Systems (IROS 2014)*; 2014.

[39] Github. *Realtime URDF filter*. Available from https://github.com/blodow/realtime_urdf_filter

[40] Fetzner A., Frese C., Frey C. 'A 3D representation of obstacles in the robots reachable area considering occlusions'. *In: ISR/Robotik 2014; 41st International Symposium on Robotics. VDE*; 2014. pp. 1–8.

[41] Sampaziotis S., Antonakoudis S., Kiatos M., *et al.* A lightweight method for detecting dynamic target occlusions by the robot body. Available from https://arxiv.org/abs/2302.06890

[42] Villani V., Pini F., Leali F., Secchi C. 'Survey on human-robot collaboration in industrial settings: safety, intuitive interfaces and applications'. *Mechatronics*. 2018, vol. 55, pp. 248–66.

[43] Arents J., Abolins V., Judvaitis J., Vismanis O., Oraby A., Ozols K. 'Human-robot collaboration trends and safety aspects: a systematic review'. *Journal of Sensor and Actuator Networks*. 2021, vol. 10(3), p. 48.

[44] Stavridis S., Papageorgiou D., Doulgeri Z. 'Dynamical system based robotic motion generation with obstacle avoidance'. *IEEE Robotics and Automation Letters*. 2017, vol. 2(2), pp. 712–18.

[45] Kouris A., Dimeas F., Aspragathos N. 'Contact distinction in human-robot cooperation with admittance control'. *2016 IEEE International Conference on Systems, Man, and Cybernetics (SMC). IEEE*; 2016. pp. 001951–56.

[46] Deniša M.M., Petrič T., Gams A, *et al.* 'A review of complaint movement Primitives'. *Robot Control*. 2016, pp. 1–17.

[47] Rosenstrauch M.J., Pannen T.J., Krüger J. 'Human robot collaboration-using kinect v2 for ISO/TS 15066 speed and separation monitoring'. *Procedia CIRP*. 2018, vol. 76, pp. 183–86.

[48] Fang H.S., Li J., Tang H., *et al.* 'AlphaPose: whole-body regional multi-person pose estimation and tracking in real-time'. *IEEE Transactions on Pattern Analysis and Machine Intelligence*. 2022.

[49] Kastritsi T., Dimeas F., Doulgeri Z. 'Progressive automation with DMP synchronization and variable stiffness control'. *IEEE Robotics and Automation Letters*. 2018, vol. 3(4), pp. 3789–96.

[50] Siciliano B., Sciavicco L., Villani L., *et al.* 'Differential kinematics and statics'. *Robotics: Modelling, Planning and Control*. 2009, pp. 105–60.

[51] Dimeas F., Moulianitis V.C., Aspragathos N. 'Manipulator performance constraints in human-robot cooperation'. *Robotics and Computer-Integrated Manufacturing*. 2018, vol. 50, pp. 222–33.

[52] Sidiropoulos A., Doulgeri Z. *Efficient DMP generalization to time-varying targets, external signals and via-points* [*arXiv*]. 2022. Available from https://arxiv.org/abs/2212.13473

[53] Vargas-Osorio S., Zúñiga C. 'A literature review on the pallet loading problem'. *Lámpsakos*. 2016, pp. 69–80.

[54] Morgan A., Wen B., Liang J., Boularias A., Dollar A., Bekris K. 'Vision-driven compliant manipulation for reliable, high-precision assembly tasks'. *Robotics: Science and Systems (RSS)*; 2021.

[55] He K., Gkioxari G., Dollár P., Girshick R. 'Mask r-cnn'. *2017 IEEE International Conference on Computer Vision (ICCV)*; 2017. pp. 2961–69.

[56] Chandel H., Vatta S. 'Occlusion detection and handling: a review'. *International Journal of Computer Applications*. 2015, vol. 120(10).

Chapter 10

Apprentice: advances toward the development of a cobot helper in assembly lines

Yunus Terzioğlu[1], Özgür Aslan[2], Burak Bolat[2], Sera Buyukgoz[3], Zeynep Sonat Baltacı[2], Sinan Kalkan[2], and Erol Şahin[2]

10.1 Introduction

Collaborative robotic manipulators, also known as cobots, are designed to work safely alongside humans, with the objective of taking industrial automation to the next level by increasing the efficiency of a human worker. The worldwide market size for cobots is projected to grow fast, and cobots are expected to accelerate the automation of frequently changing tasks, mainly in medium and small businesses. This brings in many challenges pertaining to (i) how easily such robots can be used and how much programming and design they require, (ii) how user friendly and helpful they are, and (iii) how appealing they are as coworkers.

In this chapter, we present our vision of deploying cobots on the assembly lines toward improving the productivity of the workers. Our vision is based on the observation that human workers are superior to cobots in manipulation tasks and that carrying out the versatile manipulation tasks of an assembly-line worker at the same speed and finesse is likely to remain beyond the reach of cobots in the near future.

Specifically, we report the advances we have made within ÇIRAK and its successor KALFA (Apprentice and Journeymen in Turkish) projects, referred to as an umbrella "Apprentice Project" in the rest of the chapter.

[1]Khoury College of Computer Sciences, Northeastern University, Boston, MA, USA
[2]KOVAN Research Lab, ROMER (Center for Robotics and AI Technologies) and Department of Computer Engineering, Middle East Technical University, Turkey
[3]SoftBank Robotics Europe & Sorbonne University, Institute for Intelligent Systems and Robotics, Paris, France

10.2 The apprentice vision for human-robot collaboration

In the Apprentice Project, we aim to develop technologies toward an "apprentice cobot," an underskilled robotic helper that can track the state of the worker and the assembly process and hand over (and stow away when done) the necessary tools and the parts to the worker for the current stage of the process. Such a system would only require limited manipulation capabilities, given that it is coupled with a cognitively intelligent interaction with the human worker.

The Apprentice Project vision is presented in a trailer video (Figure 10.1), where the human assembles an IKEA chair with a cobot as his apprentice.

The realization of such a robot, among others, requires:

- *human-robot interaction* (HRI) skills with a focus on nonverbal communication,
- *perception* abilities to track the worker and the workspace,
- awareness of the *assembly process*, and
- guarantees on *safety*.

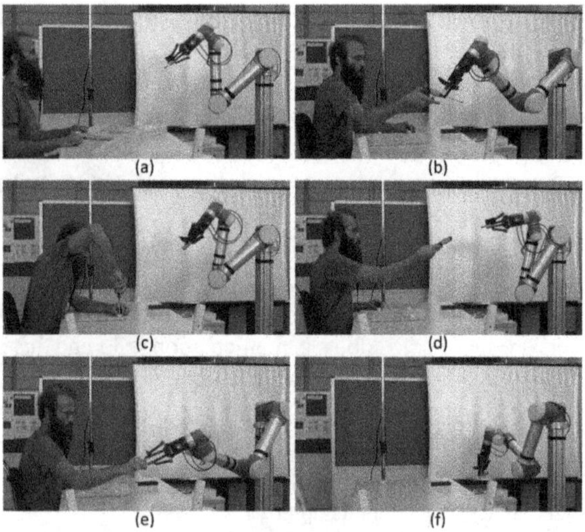

Figure 10.1 *Snapshots from the apprentice trailer video (at https://youtube/CsV363jeuJs). In the snapshots, the apprentice cobot is: (a) waiting attentively with breathing animation for the start of the assembly, (b) handing over the screwdriver to the worker, (c) attentively following the worker's assembly through mutual gaze, (d) leaning back and shaking its "head" to refuse to stow away the screwdriver because the worker still has work that he missed to finish with the screwdriver, (e) reaching out to take the screwdriver once the job involving a screwdriver is completed, and (f) stowing away the screwdriver.*

In the rest of the chapter, we will discuss some of the challenges entailing the implementation of these capabilities, review the relevant methods and approaches in the literature, and briefly report our findings. To this end, we will first describe the experimental setup which we developed and used within the project.

10.3 The apprentice setup and components

The setup, depicted in Figure 10.2, is built around a cobot, which is composed of a UR5 robotic manipulator (Universal Robots, Odense, Denmark) equipped with a Robotiq 2F-140 gripper (Robotiq Inc., Lévis, Canada); three Kinect-v2 RGB-D cameras; and three computers running robot operating system (ROS [1]) Kinetic on Ubuntu 16.04. A table is used as a shared workbench, with parts and tools placed on its sides. The cameras are used to track the pose and the gaze of the worker and the state of the assembly parts and tools used in the assembly. The three-dimensional (3D) pose information of the scene is used for collision avoidance, object detection, table top segmentation, gaze detection, and human pose detection. Each camera is connected to a separate computer (C1–C3) for distributed processing with ROS middleware, supporting the communication between different processes. C1 is used for worker gaze estimation (section 10.5.3); C2 is used for object detection (section 10.5.2) and robot control. The robot controller in C2 also handled HRIs and robot animations (section 10.4), and the main safety features (section 10.7). Finally, C3 is used for human pose estimation (section 10.5.2).

In the subsequent sections, we describe our vision and approach, along with reviews of related literature, on the four parts of research within the Apprentice Project. Toward

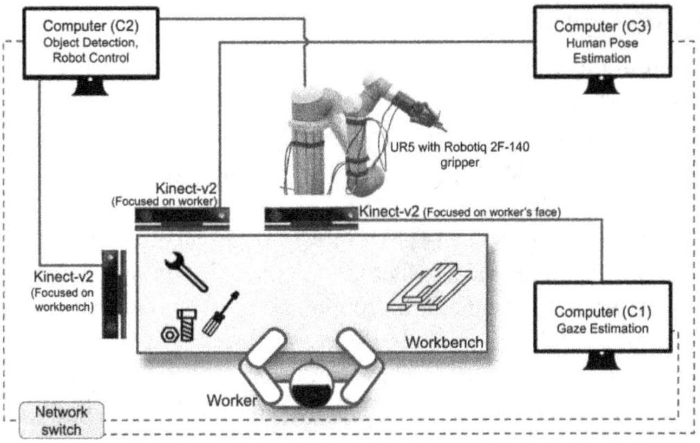

Figure 10.2 The Apprentice Project setup, which incorporates four integral components for effective human-robot collaboration: interaction, perception, assembly, and safety. The tasks underlying these components are executed in a distributed system using three computers (C1–C3) and three Kinect-v2 RGB-D cameras.

this end, we will start with the HRI part, in which we layout the challenges and our vision (section 10.4). This part will also lay the framework for other parts, namely perception (section 10.5), assembly (section 10.6), and safety (section 10.7) within our grand vision. Since the main novelty of our studies lies within our HRI vision along with this framework and the remaining three parts are relatively classical challenges, we cover these three parts in the same structure by focusing more on the challenges and related studies, with less emphasis on the work done.

10.4 Human-robot interaction using nonverbal behaviors

An essential element of a successful and long-term collaboration between humans and robots is the establishment of a simple yet effective form of HRI. Although much work has been done in developing robots with HRI capabilities, most of these platforms have not succeeded in carrying out real-world tasks. Ironically, most industrial robots, a shining example of the success of robotics in the manufacturing domain, are developed as robotic manipulators and lack any form of HRI capabilities. One notable exception was the Baxter and Sawyer cobot platforms [2] (Rethink Robotics, Massachusetts, USA), which used a screen with an animated face to communicate the robot's attention through its eyes to its users on the factory floor.

In the following subsections, we present a brief review of relevant nonverbal communication literature, primarily from the domain of social robotics; demonstrate the utility of animation principles in generating effective nonverbal interactions; and present our findings on how social aspects of an industrial robotic platform, which does not comply with the form factor norms of social robots, can be improved by animation techniques.

10.4.1 Related work

Nonverbal communication is considered to be an integral part of HRI. The significance of such information transfer between a robot and a human using social cues, such as facial expressions, gestures, and body language, has been demonstrated in several studies over the years [3]. For example, Salem *et al.* stated that arm gestures lead to a robot being interpreted as more sympathetic, lively, active, and engaged regardless of gesture congruence [4]. Arm gestures are also investigated under synchronization congruence in HRIs. Shen *et al.* have shown that coordinating robot motions with the motion characteristics of humans increases gesture recognition and task performance as well as improves the overall quality of social interaction [5]. Ende *et al.* have studied arm gestures in human interactions to replicate and evaluate them within the scope of human-robot collaboration [6]. Similarly, the utility of hand gestures such as "thumbs-up" has been adapted from human interactions to industrial domains as effective feedback mechanisms [7]. Arm motions have been specifically studied to the end of creating a formalism for legible gestures that would let the users infer a robot's goal in a variety of scenarios [8–10]. Moreover, it has been shown that incongruent and even erroneous arm gestures can increase engagement and awareness in HRIs [11, 12].

In addition to arm and hand gestures, gaze cues bear great significance in nonverbal communications, and human brain is considered to have dedicated pathways for their interpretation [13, 14]. Several studies have shown that the interaction mechanisms incorporating gaze cues can effectively be transferred into robotics domain [15–17]. Stanton and Stevens have revealed the impact of gaze to be task-dependent [18]. Rethink Robotics' Baxter and Sawyer cobot platforms incorporate a screen to animate a face, which is utilized to effectively communicate the robot's current focus on the factory floor [19] or to communicate more subtle internal states such as confusion [20].

Other studies have reported that humans intuitively synchronize their frequency of gestures to human counterparts as well as robots, drawing attention to the importance of adaptive gestures [21, 22]. Further interaction modalities, such as color, sound, and vibrations, are also studied in social robotics [23]. It also has been studied in Reference 24 that in addition to verbal and nonverbal communication channels, a backstory written for a virtual character and a social and cultural context based on this backstory can be used to create more believable characters.

Many of the animation principles as described in Reference 25 have also been used to create lifelike behaviors for robots. Some of these principles are described in detail in section 10.4.2. For example, *Anticipation* and *Secondary Action* principles have been studied in References 26 and 27 in the form of behaviors such as "sleeping." In Reference 28, Gielniak *et al.* have proposed three methods to generate *Secondary Action* cues for humanoid robots. To be able to generate more legible robot motions, Takayama *et al.* have applied the principles of *Anticipation* and *Follow Through and Overlapping Action* on a PR2 mobile robot [29]. The applications of these principles are not limited only to humanoid or collaborative robots. For example, Szafir *et al.* have implemented *Arcing*, *Anticipation*, and *Slow In and Slow Out* principles on a quadrotor to communicate its goal to the users and have shown that the utilization of these principles improved the perceived usability, safety, and naturalness of the robot's motions [30].

We build our work on this body of literature, which shows the significance of nonverbal communication of social robots for improving HRI quality in terms of fundamental metrics such as legibility, likeability, and engagement. However, industrial robots communicate a completely different set of affordances compared with their social counterparts. This is due to the fact that industrial robots are designed to optimize for work efficiency and reliability rather than human interaction performance. Following this observation, in section 10.4.3, we study how social robot behaviors can be designed for and transferred to an industrial domain without degrading overall robot platform efficiency, and in section 10.4.4, we show the (positive) effects of this transfer on HRI in a collaborative task.

10.4.2 Basic principles of animation for roboticists

The animation industry offers a significant amount of inspiration and knowledge toward imbuing industrial robot arms with HRI capabilities to transform them into working *companions*. The animators have excelled in presenting inanimate objects and rigid bodies as sentient entities to the perception of humans. They have been doing this with such a great success for such a long time that it is no longer startling for a human to create mental models of a teacup and develop empathetic attitudes toward cars, not to mention characters with more complicated kinematic chains such as the ones having anthropomorphic or zoomorphic features.

Frank Thomas and Ollie Johnston's classic book *The Illusion of Life* [25] presented "12 Principles of Animation" that were used in animating cartoons. Some of these principles, such as "Squash and Stretch," are not trivially applicable to physical entities, and some of them, such as "Slow In Slow Out," are readily inherited by robot platforms by the laws of physics and finite torque limits. In our review, we primarily focus on three principles, namely *Appeal*, *Arc*, and *Secondary Action*. These principles are not only highly applicable on a cobot platform, but they also align very well with design paradigms in the field of social robotics.

The principle of *Appeal* in animation aims at the creation of coherent and believable characters that draw the attention of the viewer and build interest. This principle helps viewers create an idea of the personality of the depicted character at a sight by representing the intended characteristics and personality of the character through its design. A similar approach to this principle has already been adopted by roboticists, and it is not a surprise (at least for an animator) that almost every social robot designed so far has face-like features, curved shells with smoothed outlines, and anthropomorphic and/or zoomorphic features, whenever applicable (Softbank Robotics' Pepper, Anki's Vector, Amazon's Astro, Mayfield Robotics' Kuri, and Fetch Mobile Manipulator). All of these can be considered successful applications of the *Appeal* principle on robots and serve the purpose of creating likeable and interesting characters for humans through their appearances.

The second principle of interest, namely *Arc*, builds on the observation that almost no organic entity in nature moves on perfectly straight trajectories and suggests the usage of smoothed curves or "Arcs" to create motion trajectories. Such curved motions generate more lifelike animations, in general. An interesting example of trajectory characteristics emerges when the animated character is a robot; however, in which case viewers praise straight-line trajectories, in particular. Humans seem to have formed a strict association between robots and straight-line motions and even happened to have invented a very peculiar dance genre following this generalization, which we believe is due for a change in the future.

Finally, the principle of *Secondary Action* in animation remarks the importance of secondary behaviors in the believability of animated characters. These behaviors, such as breathing, blinking, or shifting body weight between legs, do not seem to serve any specific purpose as the character neither relies on oxygen to survive nor needs to moisturize its eyes nor does it get tired. Yet, the presence of such actions improves the lifelikeness of the animated characters as well as highlights the events depicted in a scene.

We argue that incorporating the aforementioned principles can highly enhance the "social" aspect of a cobot and that such features can be realized with minimal hindrance to a cobot's safety, efficiency, practicality, and longevity in industrial settings.

10.4.3 Case study: interaction design for a robotic manipulator

In the Apprentice Project, we have designed and implemented a set of augmentations to improve the HRI capabilities of a robotic manipulator [31]. These augmentations were later evaluated in two user studies to demonstrate their effects on HRI using a set of 14 prevalidated HRI measures [32, 33]. These measures included constructs such as perceived intelligence, perceived sociability, social presence, and trust. Our studies suggested that the proposed augmentations significantly improved the HRI quality as perceived by the study participants working collaboratively on an assembly task with the robot. The design of these augmentations drew its primary motivation from principles commonly used in character animation (section 10.4.2), and the three principles of interest, namely *Appeal*, *Arc*, and *Secondary Action*, were realized with minimal hindrance to the manipulator's intended functionality.

The *Appeal* principle was realized by manipulating three separate cues: *Posture*, *Gaze*, and *Appearance*. In the *Posture* cue, the default idle posture of the robot was altered so as to resemble the neck of an animal from Anatidae family (e.g., a duck) in an *elbow-down* configuration. The *Gaze* cue was implemented by directing the robot's gripper toward points of interest in its configuration space, such as a toolbox that contains the necessary parts for the assembly, the human coworker's face, or the product under assembly (Figures 10.1(a), (c), and (d)). Lastly, the appearance cue was implemented by orienting the robot's gripper in a configuration such that the gripper fingers aligned vertically, resembling an animal beak/mouth, and by attaching a pair of black tinted sunglasses to create eyes. These three modifications enabled creating an animal-like general form for the robot, effectively transforming the gripper to a "head-on-a-neck" morphology as shown in Figure 10.1.

The *Arc* principle was implemented by incorporating cubic spline interpolations in the trajectory motion generation steps instead of relying on linear interpolations.

Finally, the *Secondary Action* principle was adapted to the robot arm in the form of a breathing animation. During idle times, such as when the robot was waiting for the human coworker to be done with their task, the robot utilized predefined joint configurations to compute periodic motion trajectories that resemble "breathing" behavior as can be observed in the demonstration video (link available at the caption of Figure 10.1).

10.4.4 User studies

The modifications designed for the robot were used as independent variables in two extensive user studies following a mixed-model experiment design. The findings from these studies showed that the breathing behavior (*Secondary Action*) and gaze (*Appeal*) had significant positive effects on most of the HRI measures, whereas appearance (*Appeal*) surprisingly did not have as high a significance as one would expect, and smooth motions (*Arc*) created peculiar interaction effects with appeal and secondary action factors and accumulated mixed responses from the participants. A summary of the factors' effect on our primary outcomes can be found in Reference 31.

During the experiments, we observed participants demonstrating social interaction behaviors toward the animated robot arm. These behaviors included incorporating mutual gaze with the robot even in the condition the robot didn't have "eyes"; sharing their frustration with the robot by hand gestures, smiling at, laughing with, and talking to the robot, even though the robot never talked back or showed any signs of verbal cognition; making "eye contact"; taking deep breaths alongside the robot as if they noticed that they were holding their breath out of the focus on the task at their hands and/or stress. We believe these observations alone make a valuable demonstration of how such subtle modifications on an industrial robot's "behavior" can carry out a significant outcome of triggering social behaviors in their human coworkers and promote healthy behaviors such as deep breathing that is known to reduce stress. These observations were further supported by participant comments during the exit interviews. For example, regarding the breathing behavior, several participants made comments along the lines of *"I knew that [the robot] wouldn't do something funny [to hurt me]."*

The existence of gentle breathing motions possibly helped participants read the robot's state, and the continuous nature of the motion trajectories and the periodicity of these motions seemed to be beneficial for the humans to be able to read what the robot will be doing a moment later, thus helping them feel more in control. In experimental conditions where the breathing animation was excluded from the robot behaviors, the participant comments had a totally different nature with the complaints such as *"[the robot] was staring [annoyingly] at me when I was working"* in contrast to the comments such as *"[the robot] was chilling, minding its own business while I did mine"* for the experiment condition incorporating the breathing animation. This contrast is highly intriguing as the robot's general gaze direction was identical for both conditions, and it was toward the workbench rather than being directly toward the participants.

Regarding the appearance of the robot, one participant shared their frustration as he tried to make eye contact with the robot in the condition where the robot did not have a face but a mere parallel-finger gripper that incorporated gazing behavior. He commented *"[the robot] was clearly looking at me. It felt weird to try to look back at it as I didn't know where to focus. Should I have looked at the screws by [the gripper]? I don't know, I just couldn't figure it out."* Not only such comments remark the face-like features' importance in eliminating confusions for more "organic" interactions between a human and a robot, but they also point out that a simple gazing behavior can effectively motivate a human to demonstrate active effort to make contact with a robotic arm, with no face.

Finally, regarding the arcing behavior in the form of smoothed motion trajectories, the participants have shown mixed responses as some of them found the motions "fluid" and "lifelike," several other participants shared their disapproval with the comments such as *"[the robot] felt a little too aggressive,"* and *"[the robot] resembled evil robots from <a dystopian movie>."*

In summary, as also reported in detail in Reference 31, we showed how nonverbal communication channels can be used to create an interactive robot in our study. Our results showed that principles of animation as they found life on our robot in the form of gazing, breathing, appearance, posture, and smooth motions contributed significantly to the HRI measures, such as perceived intelligence, social presence, anthropomorphism, perceived adaptability, and likeability. In an industrial perspective, these HRI-related improvements were realized with minimal to none interference with the robot's intended functionality and efficiency. Moreover, by applying these principles, we would argue that a compliant robotic manipulator's industrial efficiency can be improved as the robot is then given an intuitive, *behavior-based user interface*. Such an interface, on even the most basic level, can be useful in communicating by gazing (pointing) at work-related issues or providing a simple interface in the form of breathing that would help the workers understand whether the robot is in an error state or it is simply waiting for *something* to be done. Moreover, the robot can serve the role of a point of attention, a companion, or a source of entertainment during teatime!

10.5 Perception

Being able to perform tasks and collaborate with humans necessitates a robotic coworker to perceive its surroundings, which may include, among others, humans, their gazes and gestures, objects, parts, tools, utilities, and activities in the environment. The perception performance is exceptionally critical for a robotic coworker since it directly affects the planning and reasoning processes of the robot.

Perception systems can make use of many different sensors. An effective coverage of how perception can be performed with all these different sensors is beyond the scope of this chapter. Therefore, we limit our discussion to visual perception in this section, i.e., perception from single or a stream of RGB frames.

10.5.1 Overview of approaches

Thanks to the advances in deep learning, many problems in robotics are now solved using deep networks that are generally trained end to end using simulated and/or real data. Computer vision is no exception, and many visual perception problems are now addressed following a pipeline, as illustrated in Figure 10.3, using deep learning:

(1) *Preprocessing*: Sensors inherently provide noisy, low-level information about the environment, and the quality of perception can significantly benefit from simple "improvements" over raw information coming from the sensors. These improvements in the form of a preprocessing step can include filtering an image with a Gaussian filter to remove certain sensory artifacts or histogram equalization to improve the contrast of the image.

With the data-hungry deep learning paradigm, the nature of preprocessing has shifted significantly. In order to make a network's predictions more robust against different types of environmental challenges, such as illumination and viewpoint variations, it has become a common practice to apply various transformations on the images to mimic these variations or artifacts, and therefore, to allow the network to experience and learn from images under challenging conditions. Increasing the variety of data, these so-called *data augmentation* strategies are generally problem-specific, though the following operations are commonly used:

Figure 10.3 *Perception problems are generally addressed by following a pipeline that is composed of four steps: preprocessing, feature extraction, task-specific estimation or reasoning, and postprocessing. Blue rectangles indicate bounding boxes for predicted objects in the image.*

- geometric transformations: e.g., rotating, flipping, scaling, and shifting image
- illumination transformations: e.g., brightness, contrast, color, and saturation changes
- noise transformations: adding different types of noise, e.g., white noise, salt and pepper noise, and motion blur
- semantic transformations: changing the scene, e.g., from winter to summer; adding or removing objects

(2) *Feature extraction*: A vital step in many vision tasks is summarizing or representing the visual information in the form of a feature vector to solve the main vision task. For instance, object similarity between two objects can be determined by comparing the feature vectors of the two objects. In deep learning, since it is very difficult to design and train deep networks from scratch, a common practice is to take an existing pretrained network and use the activations of one of the last layers of the network as a feature vector. Such networks are carefully designed, tuned, and pretrained on large-scale datasets (such as ImageNet [34]), and they provide representations that can be useful in different types of problems.

These networks, referred to as *backbone networks*, include AlexNet [35], VGG [36], ResNet [37], ResNext [38], DarkNet [39], DenseNet [40], and MobileNet [41]. Important factors in choosing a backbone network are the accuracy, the memory footprint, and the running time of the model on the pretraining task. When combined with the task-specific network, the backbone network might be *frozen* and not trained, or alternatively, it might be trained further for the task at hand to yield more task-specific representations.

(3) *Task-specific estimation or reasoning*: Each robotic application has its own task-specific goal, such as detecting the manufactured product, perceiving humans

in the field, and understanding actions. For addressing such a task, a task-specific network is designed to take in the feature representation from the backbone network and find the mapping from the feature vectors to the desired task outputs, e.g., object labels, object bounding boxes, and activity labels. The nature of the task shapes the difficulty of the mapping, the architecture of the task network and how it is trained, etc. In the following subsections, we will see several examples of different vision problems.

(4) *Postprocessing*: Task networks generally make their predictions for a short period of time, generally for a single image frame. Such predictions can be significantly improved by utilizing task-level or environmental information. For example, information about the state of the task can precondition the types of objects that are going to be on the assembly table, and this can be valuable for the object detection task. As another example, by combining image-level object detections over frames of a video, detection performance can be made more robust and consistent.

10.5.2 Addressing visual detection problems

Visual detection is defined as the identification and localization of objects, tools, parts, or people in images. The coarseness of classification and localization leads to different types of detection problems: semantic segmentation, instance segmentation, panoptic segmentation, keypoint detection, and object detection. Since similar approaches are used for these problems and to keep our discussion concise, we will limit ourselves to object detection in this section. The readers are referred to References 42–44 for detailed coverage of the state of the art. In object detection, we are interested in placing rectangular bounding boxes around objects and estimating their labels.

Object detection can be generally addressed by performing classification (for object labels) and regression (for localization) for each region (i.e., box) that is likely to contain an object. Based on how the region proposals are obtained, detection methods can follow two main approaches:

Two-stage detection: In the so-called two-stage approaches, first, the region proposals in the form of bounding boxes are obtained. Then, a second stage performs classification and localization on each region proposal. The first stage is generally a network, called region proposal network (RPN), that receives feature representations of an image obtained by a backbone network and a grid of reference boxes with different ratios and sizes. For each reference box, the RPN outputs width, height, and position as well as an objectness score, which measures how likely a box contains an object. Regions (boxes) with high objectness scores are provided in the second stage. The second stage is simply a set of convolutional or dense layers mapping the features of a bounding box to its class label and fine-tuned width, height, and position. In other words, the second stage performs an additional localization step for better encapsulation of objects by boxes. Well-known two-stage detectors include Faster R-CNN [45], Cascade R-CNN [46], and Libra R-CNN [47].

One-stage detection: One-stage approaches skip the region proposal stage by placing a dense grid of potential regions of interests over the image and performing classification and localization for each region of interest in each grid cell. Well-known one-stage detectors include YOLO [48], SSD [49], and RetinaNet [50].

The absence of a region-proposal stage simplifies the design of a detector, even though this introduces training difficulties owing to a drastic imbalance between boxes overlapping with objects and boxes overlapping with backgrounds [51]. This imbalance is better handled by two-stage detectors with the help of a region proposal stage. This way, the region proposal stage can improve the performance of two-stage detectors, especially on imbalanced datasets at the expense of a computational cost.

The aforementioned one-stage and two-stage detectors rely on some reference boxes (regions) provided to the detector. Alternatively, in so-called *bottom-up* approaches, these regions and their parameters are estimated directly from visual content. This is achieved by first finding keypoints, such as top-left corner, bottom-right corner, and center of boxes, and then combining them into boxes. The advantage of such bottom-up approaches, in contrast to top-down approaches that require predesigned reference boxes, is that the detector is not limited to hand-designed reference boxes and their distributions of width, height, or position. This can be critical as hand-designed set of anchors might not match the distributions of boxes in images, and the detector might miss objects because of such mismatches.

10.5.3 Addressing facial analysis

Many human-robot collaboration scenarios can make use of rich information that can be obtained through visual analysis of a face, including recognition of the face, gender, age, expression, emotion, and gaze, which allows a robot to gain more information about and from humans. Facial analysis can be used to increase the effectiveness and naturalness of the interaction between robots and humans.

Main approaches for facial analysis follow the general pipeline outlined in Figure 10.3. The points creating the key differences between these approaches are as follows:

Preprocessing: The preprocessing step is especially crucial for facial analysis since significant portions of images in a generic setting are not faces. Therefore, facial analysis generally starts with detecting the face and aligning the face with a template (see Figure 10.4 for an example). Face detection is generally addressed with approaches similar to those used for object detection, specialized for face

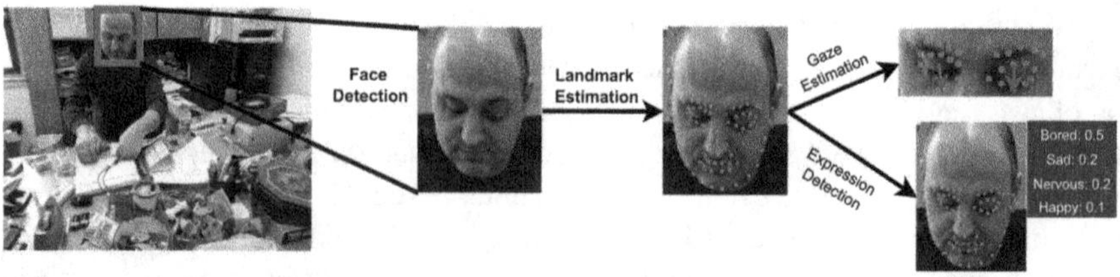

Figure 10.4 *Before performing any facial analysis, it is essential to first find the faces in an image. After this step, important landmarks are estimated as keypoints.*

images, e.g., by using human body. Face alignment is a very common preprocessing step where, first, the facial keypoints, or landmarks, are identified (Figure 10.4), and then, these landmarks are matched to a frontal face template. This matching allows correcting perspective deformation due to camera view. Moreover, these keypoints provide information about the locations of key facial features (such as eyes, eyebrows, lips, nose, etc.) and orientation of the head, and these can be valuable for several facial analysis problems.

Note that what is considered as a preprocessing step is problem-dependent and problems, such as face detection and face alignment, may be considered as the main task in certain settings.

Feature extraction: The whole face region, facial landmarks or other partitions of a face image, can be encoded using hand-designed vectors or deep networks to obtain useful features for the main facial analysis task. The current trend is to use deep learning for both representation learning (feature extraction) and main task.

Task-specific estimation or reasoning: Facial analysis can include classification problems such as recognition of a face, gender, age, emotion, and expression as well as regression problems such as gaze, emotional valence, and arousal. For both types of analyses, deep neural networks can be used.

10.5.4 Addressing activity recognition

Activity recognition is critical for collaborative robots, as robots should be able to understand and track the actions (activities) of humans in the environment. This can serve different purposes: (1) making sure that humans are doing the correct actions, (2) keeping track of the state of the task, and (3) making sure that safety regulations are not breached. With the rise of collaborative robots, activity recognition has become more important and prevalent in robotics.

Main approaches for activity recognition follow the general pipeline outlined in Figure 10.3. However, there are crucial differences compared with other visual perception problems since activity recognition relies on processing a temporal signal to perceive an inherently temporal phenomenon. How the temporal information is handled and where it is handled in the computational model lead to different types of solutions. For a review, see, e.g., Reference 52.

A promising approach is to go over the temporal axis (the frames in a video), extract features from the frames, and pass those features to a task network for predicting the label of the action. The task network is generally a recurrent architecture to capture the temporal dynamics of the action [53], even though recent years have shown convolutional architectures to be a viable alternative [54–56].

One challenge of such an approach is the length of the video signal to be processed. Videos can have different lengths, and many parts of videos may not be relevant for the recognition problem. A widely used solution is to use a sliding window over the temporal axis to process frames in a window and to make a prediction for each window [52]. Such an approach necessitates determining a suitable window size, a hop size for consecutive window positions, and mechanisms for handling overlapping and redundant predictions.

Another challenge is capturing the long-term and short-term dynamics adequately. It might be that some activities span short periods of time, whereas others might be significantly longer. Moreover, it might be that a seemingly insignificant event at the beginning of an activity needs to be linked with another event at the end for recognizing the activity. Such long-term dependencies are difficult to capture, and a multiphase approach that performs recognition at different scales might be needed [57].

A common strategy is to explicitly extract information about the temporal dynamics and provide such information as extra input to a deep network. For example, optical flow (i.e., pixel-wise movement map between consecutive frames) can be estimated using various methods, and optical flow and RGB frames can be processed by a multistream deep network – as an example study, see, e.g., Reference 55.

10.5.5 Case study: the perception system of the Apprentice Project

In the Apprentice Project, we worked on some of the perception problems in Sections 10.5.2-10.5.4 as outlined above. Specifically, we constructed a perception pipeline for detecting tools, objects, humans, and their gazes with the help of deep networks applied to RGB-D video streams.

10.5.5.1 Detection problems in the Apprentice Project

Although existing object detectors, pose estimation and human detection models, and gaze estimation networks are very capable, they are not as robust as advertised on benchmarks. Therefore, they often require tuning or usage in combination with disambiguating sources of information such as the task knowledge, any other type of contextual information, temporal consistency, etc. In addition, learning-based methods that have been trained on datasets, which do not share the characteristics of working environments that we consider, require retraining. Therefore, the collection of new labeled datasets is often necessary.

Tool detection: For a robotic coworker who is expected to perceive the tools being used and to provide the correct tools to the humans, tool detection is an essential problem. However, it is very challenging since (i) tools resemble each other, (ii) tools are generally rather small, (iii) viewpoint and illumination changes introduce significant perceptual changes, and (iv) there are no extensive datasets for tool detection.

Since the literature did not have any datasets that are suitable for training a tool detection network, as part of our project, we curated a tool detection dataset, namely ALET [58] – see Figure 10.5 for some samples. ALET dataset consists of 53 tool classes, such as meter, pen, drill, and hammer, which are frequent in a robotic setup, as well as helmet, safety eyeglasses, and gloves, which can be useful for safety reasoning (as illustrated in Figure 10.6). Each class contains 200 samples or more.

Using ALET dataset, we have trained various object detectors and obtained promising results [58]. However, our results suggested that tool detection is indeed very challenging, especially for small tools and/or cluttered environments. Before

Figure 10.5 *An important problem in robotic assembly is perceiving the environment. However, common workplaces are unstructured and pose serious challenges for perception methods (source: Reference 58).*

such a tool detector can be used reliably in practical settings, significant improvements are needed.

Human detection and pose estimation: Human and human pose detection are necessary for seamless interaction between humans and robots. Detection may provide additional safety for humans and more effective interactions with the robot. In our project, we employed a multiperson detection model, which provides key point detection, person segmentation, and pose estimation [59]. The pose estimator helps determine the state of the worker (Figure 10.7(a)). To be specific, by looking

Figure 10.6 *The labels for safety tools (helmet, eyeglasses, and gloves) in ALET dataset (source: Reference 58). These labels can be used for training deep networks for checking the safety regulations. H, helmet; M, mask; G, gloves; L, glass; D, earmuffs. The red rectangles indicate bounding boxes for the humans detected, and the connected lines within the rectangle, indicate the "skeleton" of the human within the image.*

at the detected keypoints of the human, we are able to determine the human state as "waiting," "working," and "standing up." As a result, the robot can take actions in accordance with the coworker's state.

10.5.5.2 Facial analysis problems in the Apprentice Project

We used human gaze as an important source of signal for guiding the robot coworker, e.g., to signify which object is to be worked on, what tool is to be fetched, etc. To this end, we integrated a real-time human gaze detection model that works on RGB images [60]. The detection model estimated gaze directions and head pose in 3D. The estimated 3D gaze vectors are intersected with (i) the table plane to determine which part of the table the human was looking at or (ii) a sphere positioned around the end effector of the robot to determine whether the human was looking at the robot. With this approach, we can easily calculate where the human is looking and whether or not the human and the robot coworker are sharing a gaze (Figures 10.7(b) and (c)).

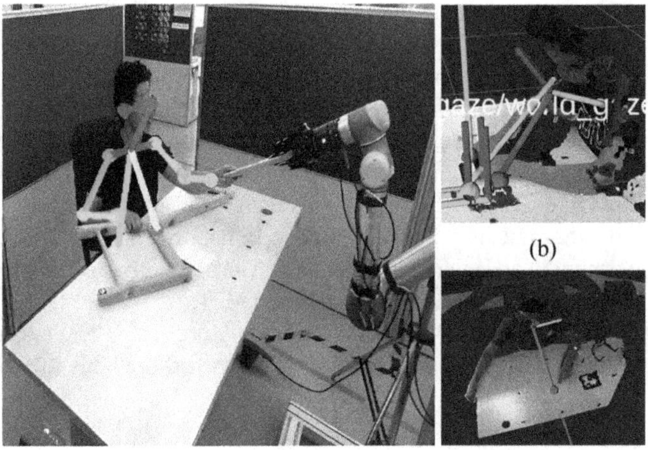

(a) (b) (c)

Figure 10.7 *In the Apprentice Project, we make use of the human pose (a) and gaze (b, c) for more effective interaction with the robot*

10.6 Assembly

Humans are remarkably talented assemblers. Since ancient times, humans used various assembled products, e.g., to hunt wild animals, to carry heavy objects, and for agriculture. With the industrial revolution, humans' assembly capacity has increased, thanks to the well-structured assembly lines, better modeling of the assemblies, and helper tools. However, today, the demands of the industry (faster production and fewer costs) exceed the capacity of human workers and autonomous nonintelligent helper robots. Therefore, there is a need for an (intelligent) assembly system that is adaptive to changes in its environment, such as workers with different skill levels, availability of the tools, and availability of the assembly components, and flexible to work on different assembly scenarios to meet the demands of the industry.

A typical assembly process can be decomposed into three main steps:

1. **Modeling:** Modeling includes the representation of the assembly structure, the object parts, connections, constraints, other relations, and features of the atomic (single) parts and subassemblies. Objects can be represented using bounding boxes, point clouds, and meshes. Selection of the object representation affects the methods used for autonomous modeling, planning, and execution. Furthermore, to elaborate the constraints of the model, computer-aided design (CAD) files can be used. These files include detailed information about geometric constraints.
2. **Planning:** Planning governs the sequence of required assembly operations leading to the final product. It also includes the motion planning for robots to execute this sequence. Each aspect of the assembly planning needs different

representations. For example, an assembly sequence can be represented using graphs, probability models, such as hidden Markov models (HMMs), or simply by a predefined computer program. For representing actions in the plan, action diagrams, functions, or relative poses of the objects and fastener information can be used. To automate assembly planning, classical search methods, learning-based methods, or controller planners can be used.

3. **Execution:** Execution is the final part of the assembly process, which includes performing the assembly individually, in collaboration with robots, or with robots working in an isolated environment. The assembly steps are tracked, and faults in these steps are detected and corrected. Tracking and fault detection is one of the most important parts of the assembly process. There exists learning-based methods and classical Markov process–based methods to automate this step.

Although these steps can be addressed separately, they are strongly interconnected. A decision in any step affects the other steps. For example, a better sequence should be found if the computed assembly sequence is too complex to execute with the available hardware. If a better sequence cannot be computed, the model of the product should be changed. This intertwined composition and complexity of each step make the assembly task difficult to manage.

10.6.1 Overview of approaches

The literature on assembly addresses the modeling, planning, and execution steps differently.

10.6.1.1 Modeling

In earlier works, CAD is used to represent the model of the assembled product. CAD files include a detailed representation of the assembly, with geometry information and related constraints on the geometries. The assembly sequence graphs, such as the AND/OR graphs, are obtained from this detailed representation, in some cases, using disassembly of the product since it reduces the complexity of finding feasible assembly sequences. Recent works use point clouds, voxels, or bounding boxes to represent the assembly geometry. Some of these works [61–63] learn the relationship (symmetry, parallelism, and contacts) between parts in the process of learning relative poses of the parts. Other works use a given relation structure (graph or tree) for generative modeling [64, 65].

In recent works, the graph-like structure of part relations within an assembly is used for learning generative networks for modeling [61, 63–69]. For example, Li *et al.* [70] proposed two network modules to extract information from the image of an assembled furniture and part point clouds. Moreover, Huang *et al.* [61] introduced a dynamic graph learning framework to make predictions only from the part point clouds. In a similar line of approach, Li *et al.* [62] learned relationships to predict the position and the scale of parts instead of their poses (position and orientation). Although recent works do not use detailed object model information, they introduce complex loss functions to assess implicitly learned relations, and they need large shape datasets to learn generalized relations.

10.6.1.2 Planning

Assembly planning is a long studied problem, emphasizing assembly sequencing [71]. There are many different ways of representing the assembly sequence. For example, de Mello and Senderson [72] propose ANDOR graphs, a simple way to represent possible sequences to obtain the assembled product. Van Holland *et al.* [73] propose a feature-based object-oriented model for assembly modeling and planning. This model combines geometric constraints, contact constraints, and object manipulation (handling) constraints to achieve an efficient representation.

Recent works [74–78] use deep reinforcement learning (DRL) to address assembly planning and execution. For example, Inoue *et al.* [74] use DRL to train a recurrent neural network to solve a high-precision peg-in-a-hole task, which cannot be accomplished with standard robot force controllers. Furthermore, due to the complexity of the assembly task, RL (reinforcement learning) algorithms can get stuck at local minima and yield suboptimal controllers. To overcome this issue, recent studies [75, 76] propose guiding RL with additional information. For instance, Thomas *et al.* [75] use CAD data to extract a geometric motion plan and a reward function that track the motion plan. Luo *et al.* [76] utilize the force information obtained by the robot's interaction with the environment. Shao *et al.* [78] use DRL to teach a robot how to manipulate the environment, which restricts the outcomes of the robot actions. That way, it helps to reduce the uncertainty of the action outcomes, and it is easier to learn manipulation skills.

10.6.1.3 Execution

After computing an optimal assembly sequence and devising a controller to execute the assembly (with or without human workers), another critical part of the assembly process is tracking the assembly and detecting and correcting faults. Even if humans are great at detecting errors, they are not efficient and reliable for large-scale and/or rapid assemblies. Thanks to the perception and identification systems, it is easier to detect faults consistently and efficiently. Combining these systems with HMM obtained with the execution of a task, the erroneous stages can be detected and reverted. Moreover, learning-based methods, such as long short-term memory, can be used to detect erroneous steps [79].

10.6.2 Case study: the assembly system of the Apprentice Project

As described above, assembly process contains complex steps with an intertwined structure. Especially, the planning of an assembly is a labor-intensive, tedious task. Therefore, in line with the apprentice vision (section 10.2), we work on assembly learning where an agent learns the correct sequence of the parts, tool usage, robot controls, and reasoning about the faults. In Figure 10.8, an overview of our approach to the assembly task is given.

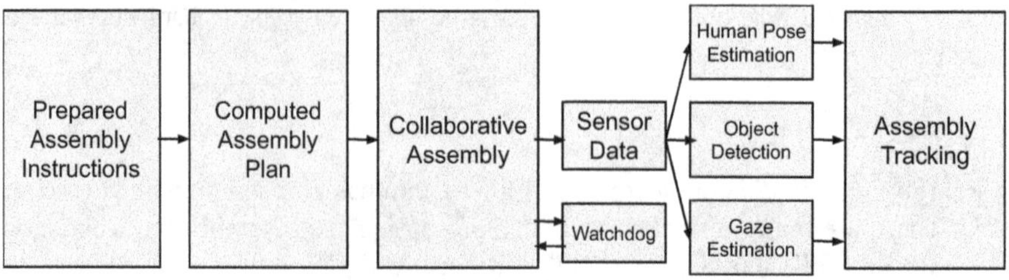

Figure 10.8 Assembly pipeline in the Apprentice Project. Predetermined assembly instructions are used to compute an assembly plan. While this assembly plan is being executed with a human worker, the assembly is tracked with the sensors and a variety of estimation systems. The safety of the human worker is monitored with a Watchdog (section 10.7).

10.6.2.1 Representing and tracking the assembly sequence

For the cobot to perceive and follow the assembly, we manually prepared a precise assembly plan for step-by-step actions. The plan included parts, tools, and details like which tools should be used on which parts and at which step of the assembly sequence. This way, the data extracted from the Kinect cameras and the AprilTag visual marker system [80] are used to track the assembly using this step-by-step plan.

10.6.2.2 Tracking the human worker

With the human pose detection systems (section 10.5.5.1), the human worker's pose and current state are tracked. The cobot uses this information to determine when to help the worker and infer if an assembly step is finished. Additionally, the worker's gaze is extracted and used in the inference process to strengthen the communication between the robot and the human worker.

10.6.2.3 Detecting errors and informing the worker

If the human worker makes a mistake, the apprentice detects the error, thanks to the detailed assembly plan and the tracking systems. The assembly plan contains furniture parts, connectors, and target subassembly information for each assembly step. Parts and their connectors collected in boxes are detected using the AprilTag visual markers, and tools are detected using a deep learning-based object detector. The objects are compared against the assembly step information to detect the errors. Additionally, human worker actions (screwing, hammering, or attaching) can be detected using learning-based action detection systems, and this information can be used to detect wrong assembly actions. If an error is detected, the apprentice informs the worker by "head" movements. These movements are achieved using the UR5's wrist joints. Moreover, the cobot does not continue with the execution of the next step, implying there is an error that needs to be fixed.

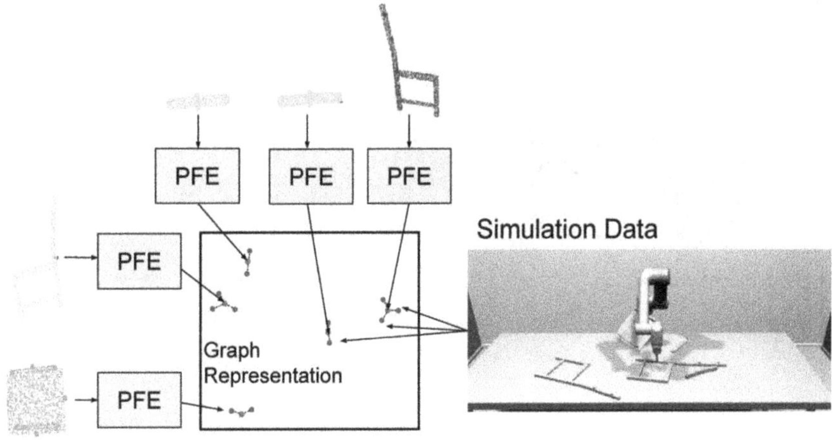

Figure 10.9 *Graph representation of the assembly. Red nodes represent the furniture parts, while blue nodes correspond to the connection sites. For the features of the part nodes, a point cloud feature extractor (PFE) is used similar to Reference 61. For the features of the connection site nodes, information from the simulation environment can be used.*

10.6.2.4 Representation of assembly sequence

To enhance the understanding of the assembly task by the cobot, we proposed to learn the assembly plan using DRL by interacting with the parts and the tools within the cobot's simulation environment, thus facilitating the use of cobot in assembly scenarios by people with little technical skills.

Similar to previous works [63, 69], point clouds are used to represent the assembly parts. We assume that only the part point clouds are available as the input, and the assembled product's point clouds are given as the target. The assembly state is represented with a graph, where each node is used to represent a part or a part's connection site (Figure 10.9). At the beginning of the assembly, only parts and their related connection site nodes have edges between them. With the execution of the assembly steps, new edges are added for each connection between the parts.

10.6.2.5 Reinforcement learning for learning the assembly sequence

We used DRL framework to learn the assembly sequence, where the agent observes the graph representation of the assembly state and determines an action (Figure 10.10; [81]). The actions can be continuous, such as relative poses of parts, or discrete, such as connecting parts from their relative connection sites. A reward signal is given to the agent (robot), depending on the similarity between the point cloud of the product assembled by the agent and the point cloud of the groundtruth assembly. In the Apprentice Project, we use the iterative closest point algorithm [82] to register these point clouds and then compute the distance between the points

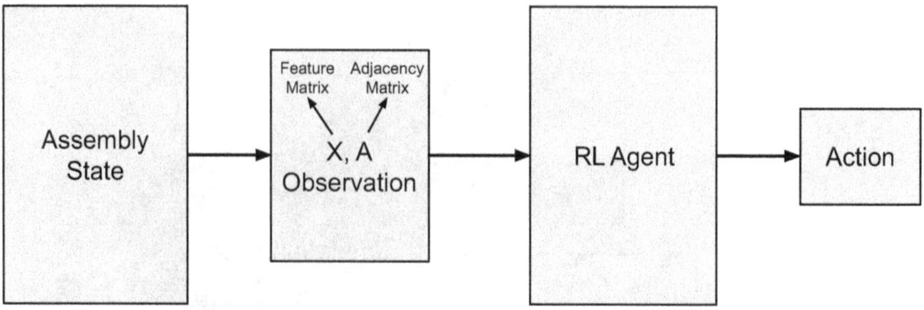

Figure 10.10 The Apprentice Project reinforcement learning pipeline [81]. The agent observes the state of the assembly (Figure 10.9) and decides a connection action.

in them. This distance is used to define the reward function. This type of reward function is used since we assumed that only the point clouds for the parts and the groundtruth product are known.

10.6.2.6 Determining error sources using a causality model

Error during the assembly execution can be detected by comparing the current state of the assembly with a predefined step-by-step plan. However, using such approach, it is not possible to detect the sources of the errors nor the steps that should be taken in order not to repeat the errors. Thus, we recommend learning a causal model in the simulation environment from the interplay between parts, tools, factors, and assembly stages and using this causal model to determine the sources of errors when an anomaly is detected.

10.6.2.7 Simulating assembly

The usage of the DRL framework mentioned in the above sections can be accomplished by using the point cloud representations only, without the need for a simulation environment. However, a lot more information can be extracted using a simulation environment. This information includes constraints induced by part geometries, connection sites, and materials of the parts, tools, and the cobot's capabilities. There exist popular physics engines [83, 84], which are extensively used in reinforcement learning and robotics research. There are also several simulation frameworks for robotic assembly [85–87], where DRL is used to learn complex manipulation and assembly tasks. In our work, we use IKEA furniture environment [87] to simulate a UR5 robot, robot's workspace, and the IKEA furniture models (see Figure 10.11 for a snapshot). The DRL framework is the same as in Figure 10.10; however, with this setup, both states and the actions are affected by the constraints. Moreover, contact forces between objects and the joint forces of the cobot can be used in the graph representation to improve learning. These pieces of information can also be used in causal model learning, improving error detection. All in all, many tools and algorithms are at play for a collaborative assembly with

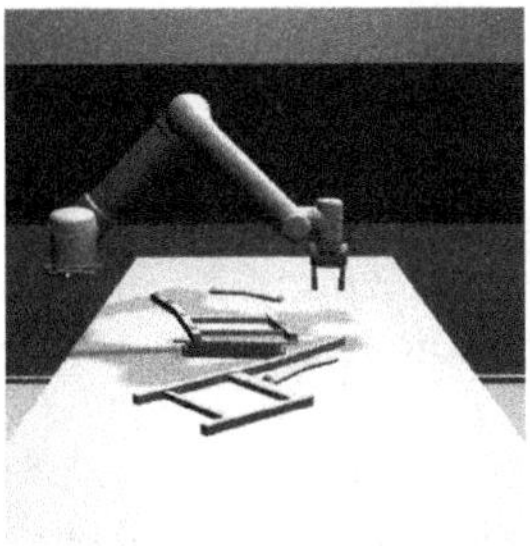

Figure 10.11 *UR5 and an IKEA chair in the IKEA simulation environment [87]. The data collected from the environment are used to learn an assembly policy.*

an intelligent robotic apprentice. When data from different sensors are involved to track the worker and the assembly status, working with a predetermined and precise assembly plan simplifies the execution of the collaborative work to a more practical level. Therefore, in the first part of our project, we worked on tracking the assembly and interacting with the human worker using a predetermined plan. However, with a predetermined plan, it is harder for the apprentice cobot to adapt to unexpected changes in the assembly. Thus, as a next step, we experiment with different ways to represent an assembly sequence and use this representation to teach a cobot agent to learn the correct assembly steps by utilizing RL algorithms and simulation environments. Moreover, our further work focuses on causal learning to improve the cobot capabilities for tracking the assembly. This way, the apprentice cobot can adapt to changes in the assembly and reason about human worker's actions resulting in better overall interaction and collaboration.

10.7 Safety

In collaborative tasks, the safety of a human is another important challenge of HRI. Safety is considered under two categories: physical safety caused by physical contact and psychological safety caused by psychological discomfort or stress induced by robots [88]. We focused on physical safety in our studies. Cobots – robotic manipulators designed to work safely alongside humans – are required to abide by certain safety measures to ensure the human safety. In this context, human vulnerability in HRI has gained significant importance as a research topic.

International Organization for Standardization (ISO) provides a wide range of classification systems, depending on the source of the hazard; however, these systems do not include the security scopes of HRI. ISO/TS 15066, published in 2016, complements ISO 10218-1/2 standards released in 2011, which are much more restrictive in terms of HRI and freedom of robot movement. In the new standards [89], four interaction operations are defined:

1. **Safety-rated monitored stop:** The robot is stopped before a robot operator enters the robot area and continues to work after the operator leaves the area.
2. **Hand guiding:** A robot operator guides the robot manually. Except for the commands given by the operator, the robot is motionless.
3. **Speed and separation monitoring:** The distance between the robot and the operator is constantly monitored and the robot slows down or stops when certain distance limits are exceeded.
4. **Power and force limiting by inherent control design:** The safety of physical contact between the operator and the robot or an outer part is aimed to be achieved by power and force limits imposed on the robot system.

In our work, we do not aim to meet ISO standards. Instead, we aim to present how ISO standards could be applied in a developing system to ensure certain safety levels are maintained.

10.7.1 Related work

Physical safety elements are evaluated under three main headings in HRI, namely crash safety, active safety, and adaptive safety [90].

- **Crash safety** ensures safe (or controlled) collisions between the human and the robot by limiting robot force/power. An example of such safety feature is the spring-based joint mechanisms used in Reference 91. In spring-based mechanisms, the amount of energy that the robot can transfer to the human in the event of a collision is limited via control laws by monitoring and protecting the system's total energy. Similar applications are used, especially in production networks, to mechanically control the forces on the robot's end effector to stay below a certain threshold [92].
- **Active safety** monitors the interaction and stops the robot's operation before a collision between the robot and the human takes place. Real-time monitoring is achievable by cameras and similar sensors. With an active control system, instead of physically separating the robot's workspace and the shared workspace with fences, actively dividing the workspace into subspaces is envisioned. In the studies, the robots, the operators and the shared workspaces are surrounded by invisible fences. These fences are controlled and managed using auxiliary systems, such as light curtains, projectors, and cameras [93]. Widely used camera-based systems ([94]) detect surrounding objects and check whether the boundaries of the safe zone are violated. While most studies consider it appropriate to follow the safe zone and stop the robot when these zones are breached, other studies focused on maximizing the efficiency in HRI and preserving the motion continuity of the robot. The studies carried out in this direction aim to maintain safety with specialized controllers instead of stopping the robot directly when the distance between the human and the robot falls below a threshold value. This type of controller, which is based on tracking the relationship between the human, the robot, and the robot's velocities, has

been studied in Reference 95. Using this method, it is guaranteed that the robot does not harm the human even in the event of an undesirably close proximity by controlling the maximum allowed speed for the robot and by ensuring that the distance between the human and the robot is greater than the distance the robot can travel in a short duration at all times. In this work, the environment is observed with cameras to avoid collisions between robots, objects, and people, which are modeled using simple geometric shapes [96].

- **Adaptive safety** monitors the human and intervenes with the robot's operation to avoid collision, without stopping the robot completely. While the robot's motion trajectory is being planned, the human's position and intentions are taken into account to proactively avoid collisions. The humans' intentions are typically determined by vision-based systems conjugated with machine learning algorithms such as context-aware pose estimation [97]. Another approach uses ECG (electrocardiogram) signals to recognize and classify the intention to move [98]. Once a future human position is estimated through intentions, an alternative trajectory is generated for the robot motion to prevent collisions.

10.7.2 Case study: an ROS-based independent Security Watchdog

In order to ensure the safety of a human operator, we have implemented an ROS-based *Security Watchdog* that is placed between the robot driver and the UR5 robot (Figure 10.12). The Security Watchdog node (Figure 10.13) checked the desired pose and velocity commands against predefined constraints to ensure the operation of the cobot remained within the desired volume, excluding the operators' body (but not the arms) and within predefined velocity limits. We focused on the two main safety element headings, namely *crash safety and active* safety, that are directly linked with our goals in the Apprentice Project. In this project, we use a cobot as a

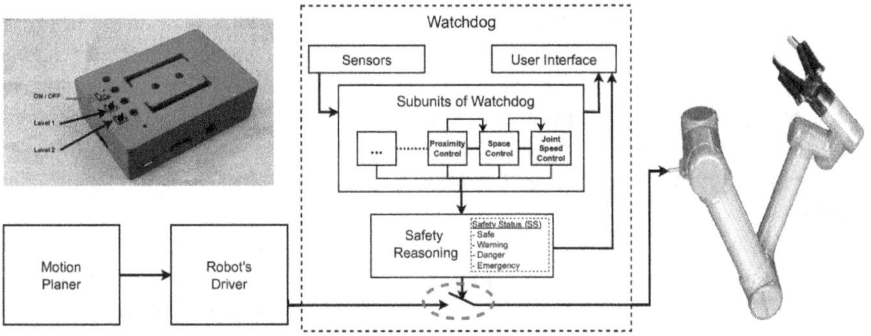

Figure 10.12 Representation of the Watchdog's (shaded box) functional positioning in a typical robot manipulation architecture. The otherwise direct connection between the robot driver and the robot is regulated by the Security Watchdog in our settings.

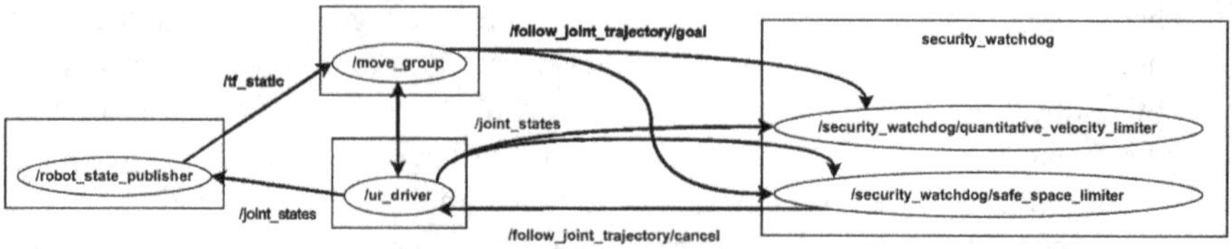

Figure 10.13 ROS node diagram of Security Watchdog

helper for the operator in a well-defined environment. In such an environment, we can meet the safety requirements we seek for the project prototype.

10.7.2.1 Architecture of the ROS Watchdog node

In the Apprentice project, we designed and fabricated an independent embedded device to execute the Watchdog software (Figure 10.12 top-left). The Watchdog ROS node in this device was placed between the robot and the robot driver, and it provided a safety layer for human operators. The Watchdog has a structure that is divided into subunits and works between the robot's driver and the robot. Figure 10.12 shows the functional deployment of the Watchdog along with its modules.

The first layer (Level 1) is responsible for the robot joint speed control, and it aims to decrease the impact of an unavoidable crash at a nonlife-threatening level to the human operator. At this layer, the robot's safe behavior is provided by keeping the robot's joint speed below predefined limits. Since keeping the robot too slow at all times is not feasible as it reduces the overall interaction and collaboration efficiency, the levels of these limits are decided by another module, namely the *safety reasoning module.*

The second layer (Level 2) aims to ensure that the robot works in accordance with the rules of three different workspace definitions. These spaces, separated by virtual fences, are: (a) *shared*, (b) *robot-only*, and (c) *human-only*. Limits are defined according to the spaces that the robot present. The robot could act with faster and more active movement in the robot-only space. The robot should act within safe collision limits in common space. Safe and efficient cooperation is aimed by ensuring that robot remains outside the space reserved for human in the human-only space. Similar limits can also be defined by using a custom motion planner. However, this approach was not preferred because it is prone to errors and does not provide the modularity and applicability as expected from the Watchdog. The Watchdog is also designed to prevent mistakes that the motion planner can make.

As shown in Figure 10.12, the Watchdog is designed to be extendable. Any number of layers can be added above the first two critical layers of safety. An example of these layers is a proximity control layer, which is planned to track the real-time position of the human(s) in the workspace with the help of an RGB-D camera system.

Finally, the outputs of the first two layers and other lower safety layers are relayed to the Safety Reasoning Module to compute a *Safety Status* value. This value serves as the final output of the Watchdog. Safety status is designed to take four different values according to the current safety assessment for the apprentice system:

- **Safe:** This is the expected operating state of the robot system. In this case, the Watchdog does not interfere in the system.
- **Warning:** This is the situation where the robot system is detected to get closed to the predefined safety limits. In this case, the Watchdog tends to reduce the operating speed of the system and gives a warning signal to the operator.
- **Danger:** This is the situation where the robot system goes out of the predefined safety limits. In this case, the movement of the robot is stopped, and the robot waits passively until it is restarted by an operator.
- **Emergency:** The result of any unexpected event that results in an unexpected conclusion by the safety reasoning module. The robot waits passively until it is restarted by an operator.

The physical human interface device (shown in Figure 10.12) is designed to establish the communication between the Watchdog and the human operator. This device informs the operator about the safety status of the system in real time. The operator is provided auditory information about the safety status of the system as well as visual guidance about the workspaces through projections on the floor.

10.7.2.2 Technical details

All modules of the Watchdog are implemented and integrated through ROS. To control the robot during tests, MoveIt! [99] motion planner framework was used. MoveIt! provides resources for motion planning, manipulation, 3D sensing, kinematics, control, and navigation applications via ROS. It is highly popular around the world as it is open source and provides models for more than 65 robots.

In autonomous robotics studies, testing development code on a real robot requires great care. Both errors in the code and problems in other libraries used may cause the robot to act unexpectedly, which may harm itself and its surroundings.

The ROS Watchdog node is designed as a layered ROS node that works independently on separate hardware, Raspberry Pi. Raspberry Pi is a single-board, Linux-based computer that can run the Watchdog module independently. The hardware is presented in Figure 10.12, which has a button for starting the safety module and registering different safety levels. It aims to prevent software- and hardware-related problems due to the "experimental code" from being transferred to the robot directly.

10.7.3 Pilot study

The pilot tests of the system were conducted by using the Moveit! planner in ROS environment. Security Watchdog was placed between UR5 robot and robot driver. The robot driver receives messages from the MoveIt! and guides the UR5 in line with these messages. The Watchdog watches these messages, if necessary, takes the necessary intervention. The flow of these messages is shown in the ROS node diagram in Figure 10.13.

In the first layer of the Watchdog, joint speed information sent by the robot is monitored. It is continuously monitored whether the robot complies with the joint speed limits. These limits are stored in an XML (Extensible Markup Language) file, which can be easily modified by the user, considering that different values may be required for different tasks. In case of a limit violation, the Watchdog stops the robot's movement using the command interface of the driver. As long as the transmitted messages are within the speed limits, the robot is free to continue executing the tasks sent through MoveIt!.

In the second layer of the Watchdog, joint angle information is continuously monitored. The forward kinematics model of the robot is provided as input to this level instead of being hard-coded. This allows being able to adapt the system to different robot platforms with relative ease. Basic geometrical shapes of spheres and planes are used to simplify the spatial boundaries around the robot joints, which are at the risk of exiting spatial boundaries. These planes and spheres are designed in a way that can be easily modified by the user, so that the module can be used for different applications and platforms, as in the first layer. The geometrical shapes used to define the physical outline of the robot are then continuously monitored to check which work zones the robot currently breaches and whether this is allowed by the rules of the current task and the state. Some demonstrations of the operation of this layer are given in Figures 10.14(a)–(c).

10.8 Challenges and open issues

10.8.1 Robustness-related challenges

The perception system is expected to function reliably – being robust against many conditions and variations in the environment. Some of these include changes in viewpoint, illumination, and background and challenges owing to variations between examples of the same class of objects, objects being deformable, and objects occluding each other. Although the literature has studied these problems extensively, proposing many different solutions, addressing these problems effectively and adequately is still an important issue.

10.8.2 Challenges with the data

With the advances in deep learning, many perception problems are now addressed using data-hungry learning-based approaches. Such approaches bring in challenges inherent in the data, e.g.,

Data scarcity: A common challenge with the data-driven deep learning approaches is the lack of data. In general, collecting and annotating large amounts of data required for training deep networks are very expensive; and therefore, data-efficient strategies are necessary. Although the literature has paid increasing attention on this challenge using transfer learning and few-shot learning strategies, this is still an unsolved problem for the community.

Data imbalance: A challenge that is closely related to data scarcity is the absence of sufficient number of samples for certain classes or tasks. Such an imbalance leads to suboptimal performance in a deep network as standard approaches favor learning of overrepresented classes or tasks. Despite various promising

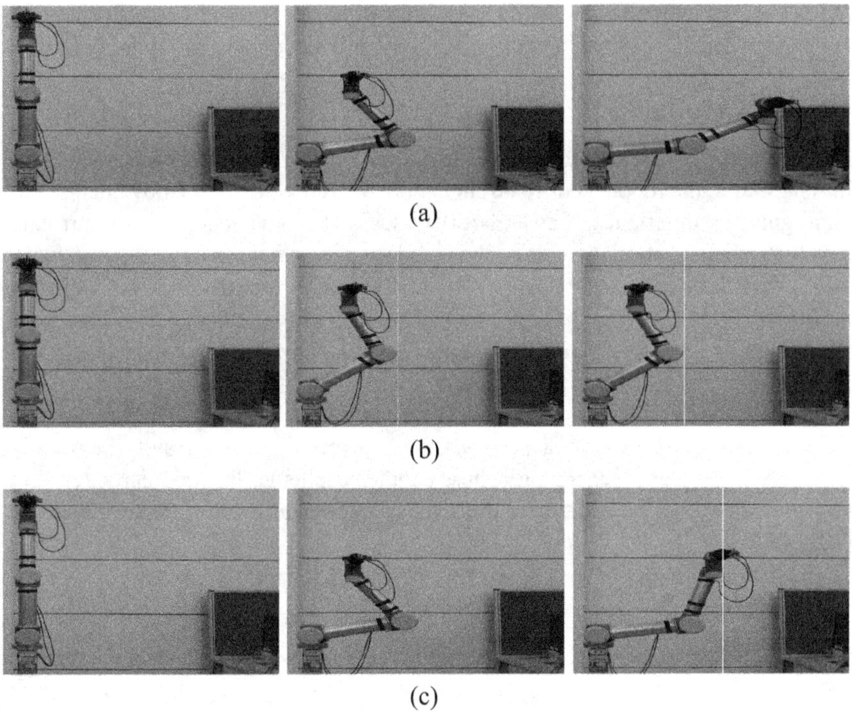

Figure 10.14 *Some example frames demonstrating the basic operation of the Security Watchdog module. (a) The robot runs without Security Watchdog. (b) The robot is working with Security Watchdog. The elbow joint crossed the secure zone boundary (yellow line). The robot's motion stops in the second picture. (c) The robot is working with Security Watchdog. The end effector has crossed the secure zone boundary (yellow line). The robot's motion stops in third picture.*

solutions using sampling, loss weighting, or postprocessing [51], mitigating data imbalance is an active research area.

Data bias: A crucial challenge with drastic social implications is bias against certain demographic groups while making predictions – e.g., it is known that facial analysis solutions for emotion estimation, face recognition, etc. perform significantly worse for certain races [100]. Making deep networks fair against different demographic groups is an active challenge and essential for widespread use of robots in human populated environments.

10.8.3 Resource-related challenges

Despite their unprecedented successes in many domains, deep learning approaches are known to be resource hungry, which can be a handicap on their widespread use in HRI settings. In HRI applications, the following resources are crucial and a bottleneck for deep learning approaches: memory, inference time, and training time.

10.8.4 Safety-related challenges

Ensuring safety according to the ISO standards brings lots of challenges alongside it. As a starter, the robot needs to be sure about the reliability of its movements. This highlights the importance of awareness of itself through detection and its surroundings through perception. Also, the robot needs to perceive the occupied space, which varies due to the movements of the operator, the parts, and the tools and adjust its motions in real time. The operator's movement also adds another layer of perception complexity to forecasting what the operator tries to achieve in the future to align the planning with respect to the operator's motion.

In addition, modularity of safety systems significantly affects their flexibility for different tasks [101]. Technology is advancing every day. As a result, the robotics hardware is frequently changing. Thus, another challenge for modular safety systems includes efficient adaption to this ever-changing hardware/software ecosystem [102].

10.8.5 Open issues

Although recent years have witnessed significant improvements in visual perception problems, there are still many open issues, especially HRI-related perception problems, in addition to the challenges outlined above:

Causality: An important concept that has not been sufficiently addressed in HRI is reasoning about the causes and effects of phenomena. For example, during the assembly of a product, an error might have occurred, and knowing the potential causes can be very useful for the human coworker. This can be addressed by taking the actions performed by the human or the robot into account and representing cause–effect relations between different states of the assembly. This can be possible by using a so-called structural causal model [103], which can also provide suggestions for correcting an error by identifying a path, in the space of cause–effect relations, leading to a "successful finish" state.

Explainability: Widespread use of robots requires that robots can explain and justify their decisions. Data-driven deep learning methods that just find a mapping between provided input–output pairs are prone to perturbations in data. An important issue is to endow the robotic coworker with mechanisms that allow the robot to explain, e.g., why it is bringing an seemingly out-of-turn item to the table or removing an item from the table that is still required.

Life-long learning: It is neither possible nor practical to foresee all possible situations a robotic coworker will face after its deployment in assembly lines; therefore, we expect a robot to be able to improve itself over its lifetime with each new experience. In other words, robots should be integrated with life-long or continual learning capabilities by design [104].

Simulation to real-world transfer: Most of the reinforcement learning agents are trained in a simulation environment and evaluated in the real world. Transferring what has been learned in a simulated environment to the real world is a challenging problem since there is a large gap between the distribution of data in simulated environments and that of the real world. Moreover, it is impossible to simulate all events that can happen in a real-world scenario.

Multiagent assembly: In collaborative tasks, a human worker model can be added to the simulation environment, or data from real-world collaborative work can be used in the learning framework to learn collaborative policies, where the robot agent does not do all the work. This way, the cobot agent can also learn how to respond to the actions taken by the human worker. Similarly, multiagent settings can be studied for cases, where each agent can learn a different step of the assembly process or collaborate on one of the steps.

Multifaceted nature of perception: Visual perception is an essential capability, even though it is not sufficient for perceiving the environment comprehensively. It needs to be integrated with the processing of other modalities (sound, tactile, etc.). However, this alone, in general, is not sufficient since perception is ill-posed, which can be addressed by utilizing background or prior information about the environment and the task.

10.9 Conclusion

In this chapter, we presented the Apprentice Project vision to advance the use of cobots in collaborative assembly tasks. Based on this vision, we have listed some of the main challenges in HRI, perception, assembly, and safety aspects. We then reviewed the related work and technologies that are appropriate to tackle these challenges and summarized the work done and our results.

Acknowledgment

This work is partially supported by TÜBİTAK through projects "ÇIRAK: Compliant Robot Manipulator Support for Assembly Workers in Factories" (117E002) "KALFA: New Methods for Assembly Scenarios with Collaborative Robots" (120E269).

References

[1] Quigley M., Conley K., Gerkey B., *et al.* 'ROS: an open-source robot operating system'. *IEEE International Conference on Robotics and Automation*; 2009.

[2] Porfirio D., Sauppé A., Albarghouthi A., Mutlu B. 'Authoring and verifying human-robot interactions'. *Annual ACM Symposium on User Interface Software and Technology*; 2018. pp. 75–86.

[3] Saunderson S., Nejat G. 'How robots influence humans: a survey of non-verbal communication in social human-robot interaction'. *International Journal of Social Robotics*. 2019, vol. 11(4), pp. 575–608.

[4] Salem M., Rohlfing K., Kopp S., Joublin F. 'A friendly gesture: investigating the effect of multimodal robot behavior in human-robot interaction'. *IEEE International Conference on Robot and Human Interactive Communication (Ro-Man)*; 2011. pp. 247–52.

[5] Shen Q., Dautenhahn K., Saunders J., Kose H. 'Can real-time, adaptive human-robot motor coordination improve humans' overall perception of a robot?'. *IEEE Transactions on Autonomous Mental Development*. 2015, vol. 7(1), pp. 52–64.

[6] Ende T., Haddadin S., Parusel S., Wusthoff T., Hassenzahl M., Albu-Schaffer A. 'A human-centered approach to robotgesture based communication within collaborative working processes'. *IEEE/RSJ International Conference on Intelligent Robots and Systems (IROS)*; 2011. pp. 3367–74.

[7] Sheikholeslami S., Moon A., Croft E.A. 'Cooperative gestures for industry: exploring the efficacy of robot hand configurations in expression of instructional gestures for human-robot interaction'. *The International Journal of Robotics Research*. 2017, vol. 36(5–7), pp. 699–720.

[8] Dragan A.D., Lee K.C.T., Srinivasa S.S. 'Legibility and predictability of robot motion'. *ACM/IEEE International Conference on Human-Robot Interaction (HRI)*; 2013. pp. 301–08.

[9] Holladay R.M., Dragan A.D., Srinivasa S.S. 'Legible robot pointing'. *International Conference on Robot and Human Interactive Communication (Ro-Man)*; 2014. pp. 217–23.

[10] Dragan A.D., Bauman S., Forlizzi J., Srinivasa S.S. 'Effects of robot motion on human-robot collaboration'. *ACM/IEEE International Conference on Human-Robot Interaction (HRI). ACM*; 2015. pp. 51–58.

[11] Salem M., Eyssel F., Rohlfing K., Kopp S., Joublin F. 'To err is human (-like): effects of robot gesture on perceived anthropomorphism and likability'. *International Journal of Social Robotics*. 2013, vol. 5(3), pp. 313–23.

[12] Riek L.D., Rabinowitch T.C., Bremner P., Pipe A.G., Fraser M., Robinson P. 'Cooperative gestures: effective signaling for humanoid robots'. *ACM/IEEE International Conference on Human-Robot Interaction (HRI)*; IEEE, 2010. pp. 61–68.

[13] Emery N.J. 'The eyes have it: the neuroethology, function and evolution of social gaze'. *Neuroscience and Biobehavioral Reviews*. 2000, vol. 24(6), pp. 581–604.

[14] Admoni H., Scassellati B. 'Social eye gaze in human-robot interaction: a review'. *Journal of Human-Robot Interaction*. 2017, vol. 6(1), pp. 25–63.

[15] Mutlu B., Kanda T., Forlizzi J., Hodgins J., Ishiguro H. 'Conversational gaze mechanisms for humanlike robots'. *ACM Transactions on Interactive Intelligent Systems (TiiS)*. 2012, vol. 1(2), pp. 1–33.

[16] Andrist S., Tan X.Z., Gleicher M., Mutlu B. 'Conversational gaze aversion for humanlike robots'. *ACM/IEEE International Conference on Human-Robot Interaction (HRI)*; 2014. pp. 25–32.

[17] Kshirsagar A., Lim M., Christian S., Hoffman G. 'Robot gaze behaviors in human-to-robot handovers'. *IEEE Robotics and Automation Letters*. 2020, vol. 5(4), pp. 6552–58.

[18] Koay K.L., Syrdal D.S., Ashgari-Oskoei M., Walters M.L., Dautenhahn K. 'Social roles and baseline proxemic preferences for a domestic service robot'. *International Journal of Social Robotics*. 2014, vol. 6(4), pp. 469–88.

[19] Sauppé A., Mutlu B. 'The social impact of a robot co-worker in industrial settings'. *ACM Conference on Human Factors in Computing Systems*; ACM, 2015. pp. 3613–22.

[20] Whitney D., Eldon M., Oberlin J., Tellex S. 'Interpreting multimodal referring expressions in real time'. *2016 IEEE International Conference on Robotics and Automation (ICRA)*; Stockholm, Sweden, 2016. pp. 3331–38.

[21] Lorenz T., Mörtl A., Hirche S. 'Movement synchronization fails during non-adaptive human-robot interaction'. *ACM/IEEE International Conference on Human-Robot Interaction (HRI)*; 2013. pp. 189–90.

[22] Ansermin E., Mostafaoui G., Sargentini X., Gaussier P. 'Unintentional entrainment effect in a context of human-robot interaction: an experimental study'. *IEEE International Conference on Robot and Human Interactive Communication*; IEEE, 2017. pp. 1108–14.

[23] Song S., Yamada S. 'Expressing emotions through color, sound, and vibration with an appearance-constrained social robot'. *ACM/IEEE International Conference on Human-Robot Interaction (HRI)*; Association for Computing Machinery, 2017. pp. 2–11.

[24] Simmons R., Makatchev M., Kirby R., *et al.* 'Believable robot characters'. *AI Magazine*. 2011, vol. 32(4), pp. 39–52.

[25] Thomas F., Johnston O., Thomas F. 'The illusion of life: Disney animation' in Hyperion, New York; 1995.

[26] Van Breemen A. 'Bringing robots to life: applying principles of animation to robots' in *Shaping Human-Robot Interaction Workshop Held at CHI*; 2004. pp. 143–44.

[27] Van Breemen A., Yan X., Meerbeek B. 'ICat: an animated user-interface robot with personality'. *International Conference on Autonomous Agents and Multi Agent Systems*; 2005. pp. 143–44.

[28] Gielniak M.J., Liu C.K., Thomaz A.L. 'Secondary action in robot motion'. *IEEE International Conference on Robot and Human Interactive Communication (RoMan)*; 2010. pp. 310–15.

[29] Takayama L., Dooley D., Ju W. 'Expressing thought: improving robot readability with animation principles'. *ACM/IEEE International Conference on Human-Robot Interaction (HRI)*; IEEE, 2011. pp. 69–76.

[30] Szafir D., Mutlu B., Fong T. 'Communication of intent in assistive free flyers'. *ACM/IEEE International Conference on Human-Robot Interaction (HRI)*; IEEE, 2014. pp. 358–65.

[31] Terzioğlu Y., Mutlu B., Şahin E. 'Designing social cues for collaborative robots: the role of gaze and breathing in human-robot collaboration'. *ACM/IEEE*

International Conference on Human-Robot Interaction (HRI); 2020. pp. 343–47.
[32] Bartneck C., Kulić D., Croft E., Zoghbi S. 'Measurement instruments for the anthropomorphism, animacy, likeability, perceived intelligence, and perceived safety of robots'. *International Journal of Social Robotics*. 2009, vol. 1(1), pp. 71–81.
[33] Heerink M., Krose B., Evers V., Wielinga B. 'Measuring acceptance of an assistivesocial robot: a suggested toolkit'. *IEEE International Conference on Robot and Human Interactive Communication (Ro-Man)*; 2020. pp. 528–33.
[34] Russakovsky O., Deng J., Su H., *et al.* 'Imagenet large scale visual recognition challenge'. *International Journal of Computer Vision*. 2015, vol. 115(3), pp. 211–52.
[35] Krizhevsky A., Sutskever I., Hinton G.E. 'Image Net classification with deep convolutional neural networks' in Pereira F., Burges C.J.C., Bottou L., *et al.* (eds.). Advances in neural information processing systems. Curran Associates; pp. 84–90.
[36] Simonyan K., Zisserman A. 'Very deep convolutional networks for large-scale image recognition'. 2015.
[37] He K., Zhang X., Ren S., Sun J. 'Deep residual learning for image recognition'. *IEEE Conference on Computer Vision and Pattern Recognition (CVPR)*; Las Vegas, NV, USA, 2016.
[38] Xie S., Girshick R., Dollar P., Tu Z., He K. 'Aggregated residual transformations for deep neural networks'. *IEEE Conference on Computer Vision and Pattern Recognition (CVPR)*; Honolulu, HI, 2017.
[39] Redmon J., Farhadi A. *YOLO9000: better, faster, stronger*; 2016.
[40] Huang G., Liu Z., Van Der Maaten L., Weinberger K.Q. 'Densely connected convolutional networks'. 2018.
[41] Howard A.G., Zhu M., Chen B., *et al.* 'MobileNets: efficient convolutional neural networks for mobile vision applications'. *CoRR*. 2017.
[42] Zhao Z.Q., Zheng P., Xu S.T., Wu X. 'Object detection with deep learning: a review'. *IEEE Transactions on Neural Networks and Learning Systems*. 2019, vol. 30(11), pp. 3212–32.
[43] Guo Y., Liu Y., Georgiou T., Lew M.S. 'A review of semantic segmentation using deep neural networks'. *International Journal of Multimedia Information Retrieval*. 2018, vol. 7(2), pp. 87–93.
[44] Elharrouss O., Al-Maadeed S., Subramanian N., *et al.* 'Panoptic Segmentation: a review'.2021.
[45] Ren S., He K., Girshick R., Sun J. 'Faster R-CNN: towards real-time object detection with region proposal networks'. *CoRR*; abs/1506.01497, 2015.
[46] Cai Z., Vasconcelos N. 'Cascade R-CNN: delving into high quality object detection'. 2017.
[47] Pang J., Chen K., Shi J., Feng H., Ouyang W., Lin D. 'Libra R-CNN: towards balanced learning for object detection'. *CoRR*, 2019.
[48] Redmon J., Divvala S., Girshick R., Farhadi A. 'You only look once: unified, real-time object detection'. 2016. pp. 779–78.

[49] Liu W., Anguelov D., Erhan D. 'SSD: single shot multibox detector'. *Lecture Notes in Computer Science*. 2016, pp. 21–37.

[50] Lin T.Y., Goyal P., Girshick R., He K., Dollar P. 'Focal loss for dense object detection'. 2017. pp. 2999–3007.

[51] Oksuz K., Cam B.C., Kalkan S., Akbas E. 'Imbalance problems in object detection: a review'. *IEEE Transactions on Pattern Analysis and Machine Intelligence*. 2020, vol. 43(10), pp. 3388–415.

[52] Rodríguez-Moreno I., Martínez-Otzeta J.M., Sierra B., Rodriguez I., Jauregi E. 'Video activity recognition: state-of-the-art'. *Sensors (Basel, Switzerland)*. 2019, vol. 19(14), 3160.

[53] Donahue J., Hendricks L.A., Guadarrama S., *et al.* 'Long-term recurrent convolutional networks for visual recognition and description'. *2015 IEEE Conference on Computer Vision and Pattern Recognition (CVPR)*; Boston, MA, USA, 2015. pp. 2625–34.

[54] Tran D., Bourdev L., Fergus R., Torresani L., Paluri M. 'Learning spatiotemporal features with 3D convolutional networks'. *2015 IEEE International Conference on Computer Vision (ICCV)*; Santiago, Chile, 2015. pp. 4489–97.

[55] Feichtenhofer C., Pinz A., Zisserman A. 'Convolutional two-stream network fusion for video action recognition'. *2016 IEEE Conference on Computer Vision and Pattern Recognition (CVPR)*; Las Vegas, NV, USA, 2016. pp. 1933–41.

[56] Kalfaoglu M.E., Kalkan S., Alatan A.A. 'Late temporal modeling in 3D cnnarchitectures with bert for action recognition'. *European Conference on Computer Vision*; 2020. pp. 731–47.

[57] Feichtenhofer C., Fan H., Malik J., He K. 'SlowFast networks for video recognition'. *IEEE/CVF International Conference on Computer Vision (ICCV)*; 2019. pp. 6202–11.

[58] Kurnaz F.C., Hocaoglu B., Yılmaz M.K., Sulo I., Kalkan S., ALET (Automated Labeling of Equipment and Tools) 'A dataset for tool detection and human worker safety detection'. *European Conference on Computer Vision*; Springer, 2020. pp. 371–86.

[59] Kocabas M., Karagoz S., Akbas E. 'Multiposenet: fast multi-person pose estimation using pose residual network'. *arXiv:1807.04067*. 2018.

[60] Fischer T., Chang H.J., Demiris Y., Demiris Y. 'RT-GENE: real-time eye gaze estimation in natural environments'. *European Conference on Computer Vision*; 2018. pp. 339–57.

[61] Huang J., Zhan G., Fan Q., *et al.* 'Generative 3D part assembly via dynamic graph learning'. *CoRR*; 2020.

[62] Li J., Niu C., Xu K. 'Learning part generation and assembly for structure-aware shape synthesis'. *Proceedings of the AAAI Conference on Artificial Intelligence*. 2020, vol. 34(7), pp. 11362–69.

[63] Harish A.N., Nagar R., Raman S. 'RGL-NET: a recurrent graph learning framework for progressive part assembly'. *CoRR*; 2021.

[64] Mo K., Guerrero P., Yi L. 'Structurenet: hierarchical graph networks for 3D shape generation'. *CoRR*; 2019.

[65] Mo K., Wang H., Yan X., Guibas L.J. 'Learning to generate 3D point Cloudshapes from part tree conditions'. *European Conference on Computer Vision*; Springer, 2020. pp. 683–701.

[66] Li J., Xu K., Chaudhuri S., Yumer E., Zhang H., Guibas L. 'ACM transactions on graphics (TOG)'. *Grass: Generative Recursive Autoencoders for Shape Structures*. 2017, vol. 36, pp. 1–14.

[67] Wu Z., Wang X., Lin D., Lischinski D., Cohen-Or D., Huang H. 'Sagnet: Structure-aware generative network for 3D-shape modeling'. *ACM Transactions on Graphics*. 2019, vol. 38(4), pp. 1–14.

[68] Shu D.W., Park S.W., Kwon J. '3D point cloud generative adversarial network based on tree structured graph convolutions'. *IEEE/CVF International Conference on Computer Vision*; 2019. pp. 3859–68.

[69] Jones R.K., Barton T., Xu X., *et al*. 'Shape assembly: Learning to generateprograms for 3D shape structure synthesis.'. *ACM Transactions on Graphics*. 2020, vol. 39(6), pp. 1–20.

[70] Li Y., Mo K., Shao L. 'Learning 3D part assembly from a single image'. *arXiv. 2020;abs/2003.09754*.

[71] Jiménez P. 'Survey on assembly sequencing: a combinatorial and geometrical perspective'. *Journal of Intelligent Manufacturing*. 2013, vol. 24(2), pp. 235–50.

[72] Homem de Mello L.S., Sanderson A.C. 'And/Or graph representation of assembly plans'. *IEEE Transactions on Robotics and Automation*. 1990, vol. 6(2), pp. 188–99.

[73] Holland W.V, Bronsvoort W.F. 'Assembly features in modeling and planning'. *Robotics and Computer-Integrated Manufacturing*. 2000, vol. 16(4), pp. 277–94.

[74] Inoue T., De Magistris G., Munawar A., Yokoya T., Tachibana R. 'Deep reinforcement learningfor high precision assembly tasks'. *IEEE/RSJ International Conference on Intelligent Robots and Systems (IROS)*; 2017. pp. 819–25.

[75] Thomas G., Chien M., Tamar A., Ojea J.A., Abbeel P. 'Learning robotic assembly from CAD'. *IEEE International Conference on Robotics and Automation (ICRA)*; IEEE, 2018. pp. 3524–31.

[76] Luo J., Solowjow E., Wen C., *et al*. 'Reinforcement learning on variable impedance controller for high-precision robotic assembly'. *International Conference on Robotics and Automation (ICRA)*; IEEE, 2019. pp. 3080–87.

[77] Fan Y., Luo J., Tomizuka M. 'A learning framework for high precision industrial assembly'. *International Conference on Robotics and Automation (ICRA)*; IEEE, 2019. pp. 811–17.

[78] Shao L., Migimatsu T., Bohg J. 'Learning to scaffold the development of robotic manipulation skills'. *IEEE International Conference on Robotics and Automation (ICRA)*; IEEE, 2020. pp. 5671–77.

[79] Altan D., Sariel S. 'What went wrong? identification of everyday object manipulation anomalies'. *Intelligent Service Robotics*. 2021, vol. 14(2), pp. 215–34.

[80] Olson E. 'A robust and flexible visual fiducial system'. *IEEE International Conference on Robotics and Automation*; IEEE, 2011. pp. 3400–07.

[81] Aslan Ö., Bolat B., Bal B., Tumer T., Sahin E., Kalkan S. 'AssembleRL: learning to assemble furniture from their point clouds'. *IEEE/RSJ International Conference on Intelligent Robots and Systems (IROS)*; 2022.

[82] Besl P.J., McKay N.D. 'A method for registration of 3-D shapes'. *IEEE Transactions on Pattern Analysis and Machine Intelligence*. 1992, vol. 14(2), pp. 239–56.

[83] Todorov E., Erez T., Tassa Y. 'Mujoco: a physics engine for model-based control'. *IEEE/RSJ International Conference on Intelligent Robots and Systems*; IEEE, 2012. pp. 5026–33.

[84] Coumans E., Bai Y. PyBullet, a Python Module for physics simulation for games, robotics and machine learning. 2016. Available from http://pybullet.org [Accessed 7 Jun 2023].

[85] Brockman G., Cheung V., Pettersson L., *et al. OpenAI Gym*. 2016.

[86] Zhu Y., Wong J., Mandlekar A., *et al.* 'Robosuite: a modular simulation framework and benchmark for robot learning'. *CoRR*. 2020.

[87] Lee Y., Hu E.S., Lim J.J. 'IKEA furniture assembly environment for long horizon complex manipulation tasks'. *IEEE International Conference on Robotics and Automation (ICRA)*; 2021. pp. 6343–49.

[88] Lasota P.A., Song T., Shah J.A. 'A survey of methods for safe human-robot interaction'. 2017.

[89] Rosenstrauch M.J., Krüger J. 'Safe human-robot-collaboration-introduction and experiment using ISO/TS 15066'. *International Conference on Control, Automation and Robotics (ICCAR)*; 2017. pp. 740–44.

[90] Michalos G., Makris S., Tsarouchi P., Guasch T., Kontovrakis D., Chryssolouris G. 'Design considerations for safe human-robot collaborative workplaces'. *Procedia CIRP*. 2015, vol. 37, pp. 248–53.

[91] Laffranchi M., Tsagarakis N.G., Caldwell D.G. 'Safe human-robot interaction via energy regulation control'. *IEEE/RSJ International Conference on Intelligent Robots and Systems*; 2009. pp. 35–41.

[92] Cherubini A., Passama R., Crosnier A., Lasnier A., Fraisse P. 'Collaborative manufacturing with physical human-robot interaction'. *Robotics and Computer-Integrated Manufacturing*. 2016, vol. 40, pp. 1–13.

[93] Vivo G., Zanella A., Tokcalar O., Michalos G. 'The ROBO-PARTNER EC project: CRF activities and automotive scenarios'. *Procedia Manufacturing*. 2017, vol. 11, pp. 364–71.

[94] Vogel C., Walter C., Elkmann N. 'Safeguarding and supporting future human-robot cooperative manufacturing processes by a projection- and camera-based technology'. *Procedia Manufacturing*. 2017, vol. 11, pp. 39–46.

[95] Zanchettin A.M., Ceriani N.M., Rocco P., Ding H., Matthias B. 'Safety in human-robot collaborative manufacturing environments: metrics and control'. *IEEE Transactions on Automation Science and Engineering*. 2016, vol. 13(2), pp. 882–93.

[96] Safeea M., Mendes N., Neto P. 'Minimum distance calculation for safe human-robot interaction'. *Procedia Manufacturing*. 2017, vol. 11, pp. 99–106.

[97] Liu H., Wang L. 'Collision-free human-robot collaboration based on context awareness'. *Robotics and Computer-Integrated Manufacturing*. 2021, vol. 67, 101997. Available from https://www.sciencedirect.com/science/article/pii/S0736584520302088

[98] Buerkle A., Eaton W., Lohse N., Bamber T., Ferreira P. 'EEG based arm movement intention recognition towards enhanced safety in symbiotic human-robot collaboration'. *Robotics and Computer-Integrated Manufacturing*. 2021, vol. 70, 102137. Available from https://www.sciencedirect.com/science/article/pii/S0736584521000223

[99] Chitta S., Sucan I., Cousins S. 'MoveIt! [ROS topics]'. *IEEE Robotics & Automation Magazine*. 2012, vol. 19(1), pp. 18–19.

[100] Cheong J., Kalkan S., Gunes H. 'The Hitchhiker's guide to bias and fairness in facial affective signal processing: overview and techniques'. *IEEE Signal Processing Magazine*. 2021, vol. 38(6), pp. 39–49.

[101] Arents J., Abolins V., Judvaitis J., Vismanis O., Oraby A., Ozols K. 'Human-robot collaboration trends and safety aspects: a systematic review'. *Journal of Sensor and Actuator Networks*. 2021, vol. 10(3), p. 48.

[102] Karnouskos S., Sinha R., Leitão P., Ribeiro L., Strasser T.I. 'The applicability of ISO/IEC 25023 measures to the integration of agents and automation systems'. *Annual Conference of the IEEE Industrial Electronics Society*; IEEE, 2018. pp. 2927–34.

[103] Pearl J. '*Causality*' in Cambridge University Press; 2009.

[104] Thrun S., Mitchell T.M. 'Lifelong robot learning'. *Robotics and Autonomous Systems*. 1995, vol. 15(1–2), pp. 25–46.

Chapter 11

Wrap up and open research directions

This book has provided an overview of the latest research and developments in the field of human-robot collaboration, with a particular focus on the potential for industrial adoption. By working together, robots and humans can take advantage of each other's strengths and complement each other's abilities. This will become increasingly important as Industry 5.0, which focuses on the cooperation between humans and machines, becomes more prevalent.

The presented chapters cover a range of topics, including teaching by demonstration, motion generation, task coordination, and safety. Novel methods are presented for tactile sensing, human detection from 3D cameras, collision avoidance, robot compliance, and safety distance separation. While some of the presented results aim to find novel ways for robotic systems to respect current safety standards, others are pushing forward toward evolving them. Despite the promising technologies being developed, there are issues of liability, certification, standardization, and human trust that need to be addressed before widespread adoption can occur. Building human trust toward a robot depends on many factors including, the human's disposition as well as their ongoing experience, which can be influenced by the robot's collaborative performance. Trust is, therefore, a dynamic process that increases as long as the robot is collaborating in a transparent and explainable manner, in the sense that the human understands what the robot is doing and for that purpose.

To widen the spectrum of applications of human-robot collaboration, further research is required into developing robots that can adapt their autonomy level based on their own understanding of human actions or intentions as well as the changes in the collaborative workspace environment. This is of particular importance in collaborative tasks in which the robot and the human are connected physically exchanging forces directly or via an object that is collaboratively manipulated. In this direction, advancements in sensor technology that can provide enhanced perception capabilities to the robotic system can extend the robot's understanding potential. However, developing learning methods that reliably infer humans' goals is still an open research field. Methods and models for online predictions of human goals and real-time adaptation are needed. Based on the confidence level of such predictions, the robot should vary its autonomy level so that it can be of real assistance to humans. The development of such predictive models is an exciting area for future work. Last, any research results should be objectively evaluated with benchmarking and validation methods that are currently missing.

As we continue to explore and develop new methods for human-robot collaboration, we will constantly unlock new potential in human-robot collaboration for making industrial applications more efficient and effective.

Index

Printed in the USA
CPSIA information can be obtained
at www.ICGtesting.com
JSHW011335040324
58545JS00003B/24

9 781839 535987